Ancestors in Our Genome

ANCESTORS IN OUR GENOME

The New Science of Human Evolution

Eugene E. Harris

OXFORD
UNIVERSITY PRESS

OXFORD

UNIVERSITY PRESS

Oxford University Press is a department of the University of Oxford.
It furthers the University's objective of excellence in research, scholarship,
and education by publishing worldwide.

Oxford   New York
Auckland   Cape Town   Dar es Salaam   Hong Kong   Karachi
Kuala Lumpur   Madrid   Melbourne   Mexico City   Nairobi
New Delhi   Shanghai   Taipei   Toronto

With offices in
Argentina   Austria   Brazil   Chile   Czech Republic   France   Greece
Guatemala   Hungary   Italy   Japan   Poland   Portugal   Singapore
South Korea   Switzerland   Thailand   Turkey   Ukraine   Vietnam

Oxford is a registered trademark of Oxford University Press
in the UK and certain other countries.

Published in the United States of America by
Oxford University Press
198 Madison Avenue, New York, NY 10016

Library of Congress Cataloging-in-Publication Data

Harris, Eugene E. (Professor), author.
Ancestors in our genome : the new science of human evolution / Eugene E. Harris.
p. ; cm.
Includes bibliographical references and index.
ISBN 978-0-19-997803-8 (alk. paper)
I. Title. [DNLM: 1. Genome, Human.   2. Biological Evolution.   3. Hominidae—genetics.
QU 460] QP624
611'.018166—dc23
2014010296

9 8 7 6 5 4 3 2 1
Printed in the United States of America
on acid-free paper

To my parents, Joan and Whitney, and my son, Bryan

CONTENTS

LIST OF FIGURES AND TABLES

FIGURES

TABLE

ACKNOWLEDGMENTS

I thank the following colleagues for their generous time in commenting on chapters: Drs. Günter Bräuer, Henry Harpending, Jody Hey, Carles Lalueza-Fox, Diogo Meyer, Maryellen Ruvolo, Caro-Beth Stewart, as well as two anonymous reviewers. Rebecca Ham deserves very special thanks for her extremely thorough editorial advice, organization suggestions, and encouragement. All illustrations were rendered with the expert and very gracious help of Kei Hayashi. Drs. Sarah Tishkoff, Hopi Hoekstra, Günter Bräuer, Michael Day, and John Fleagle kindly allowed me to use research photos of theirs. Tim White and David Brill allowed me to use photos of fossil skulls. Stephen Nash generously allowed me to use his original illustrations of papionin monkeys. I thank Jeremy Lewis, Erik Hane and Michael Stein at Oxford University Press for their generous editorial advice. An extra special thanks goes to my agent Deirdre Mullane for her sound editorial advice and her steadfast and generous support in guiding this project along. I also owe a debt of gratitude to my previous mentors Terry Harrison, Clifford Jolly, Todd Disotell and Jody Hey. I also thank Dean Falk for her helpful advice and support, Lisa Schlotterhausen for her editorial suggestions, as well as my colleagues Avelin Malyango, Varsha Pilbrow, and Scott Sherman. Any inaccuracies or mistakes in interpretation of other people's work are my own.

PROLOGUE

We are now in an era of enormous potential for studies of our evolutionary past. With the determination of the full human genome sequence in 2001—the culmination of a scientific quest begun almost fifty years earlier when James Watson and Francis Crick first discovered the molecular structure of DNA—we are at the beginning of a genomic voyage back in time. The pace at which full genomes of our primate relatives are also being sequenced is exhilarating; genomes of the common chimpanzee, bonobo, gorilla, orangutan, and macaque monkey have already been fully determined and that of other primates is well under way. Armed with sophisticated new tools, researchers are starting to examine variations in our genome among diverse peoples of the world and compare it with those of our close primate relatives in order to answer age-old questions about where, when, and how we evolved. For the first time, we are seeing our ancestors in our genome, obtaining new and often astonishing views of our evolutionary past, and we are already beginning to identify genomic features in which humans are similar to other primates and features in which humans are unique. Numerous large-scale studies have also started to catalogue millions of differences in DNA among individuals from around the world, providing us with more finely detailed knowledge of genetic diversity within our own species. Among other remarkable insights, genomic analyses are enabling us to identify with great certainty evolutionary relationships among our ape cousins; to estimate more precisely the time and nature of the evolutionary process that produced the human lineage; to identify the genetic bases of our species' adaptations, such as increased brain size and language; and to determine when and by which genetic mechanisms human populations adapted to different environments around the world. In short, the unprecedented scale of genomic evidence now being collected is revolutionizing how and what we can learn about our origins.

The source of all this information is found in the nucleotides of the human genome, the strands of molecules that join together in varying but precise ways to form the ladderlike steps of hereditary information encoded in the helical structure of our DNA. Although each person's genome is unique, reflecting the individual biological blueprints we inherit from our parents, the basic structure of the genome is similar across our species. (This is the reason why some scientists refer to *the* human genome, even though this is not technically accurate.) The entire human genome is somewhat greater than three billion nucleotides in length, organized into segments that combine to form genes that code for the proteins in the human body and influence the development of our every physical feature, from eye color to blood type. Like pearls strung along a DNA necklace, genes can be of different lengths, ranging from a few hundred nucleotides for the smallest genes to a few million nucleotides for the longest.

A gene's A, C, G, and T nucleotides provide the instructions to our body about how to construct its different proteins. Proteins, of course, are macromolecules that play innumerable important functions and are the workhorses of our bodies. A protein like collagen plays a structural role in making our bones, ligaments, and tendons strong; the protein hemoglobin in our red blood cells helps transport oxygen and carbon dioxide in and out of our tissues; and the protein fibrin is essential for normal blood clotting. Proteins themselves are actually composed of chains of little building block molecules called amino acids—of which we have twenty different types—and each different protein has a unique chain of these amino acids. The type and order of DNA nucleotides in a gene provides the instructions on how to put these amino acids together.

Perhaps surprisingly, however, genes make up only about 1.5% percent of the entire human genome. This means that only a very small fraction of our vast genome is actually coding for the proteins that carry out important structural and functional roles that make our body work. The remainder of the genome was long considered to be almost entirely "junk DNA" (i.e. non-functional), but more recent studies have estimated that as much as 4% to 12% of this remaining DNA also has important functions and perhaps plays important roles in controlling the expression of genes.[1,2] For example, many short elements within the genome have been identified that can turn genes on and off, or can increase and decrease the amounts of protein genes produce. These gene switches are responsible for activating different sets of genes in different anatomical regions of our body, which helps to explain why our brains have very different characteristics from our livers.

The sequence of a genome is determined through a complex laboratory process using increasingly efficient technologies to resolve the precise order of the nucleotide bases that pair together to form the steps of the ladder within our DNA. Since a human genome is so large, its sequence needs to be deciphered in small segments, which are then assembled together using sophisticated computer programs to produce the entire genomic sequence. One significant finding from initial analyses of our genome is that the number of genes it contains is far fewer than we had long believed. Instead of the 50,000 to 100,000 genes suggested in genetics textbooks as recently as fifteen years ago, the human genome is now considered to be made up of approximately 21,000 genes.

A FRESH LOOK AT OLD QUESTIONS ABOUT OUR PAST

But even as old mysteries are finally solved, the new DNA evidence leads us forward by causing us to revisit old and persistent questions about our evolutionary origins, as well as raising important new ones, many of which we would never have thought to ask. One of the greatest events in all of evolution—from a human-centric standpoint—is the one that led to the divergence of the human lineage from our great ape cousins—the chimpanzees, gorillas, and orangutans. Within these pages, I'll tell the story of the genetic quest, from small stretches of DNA to entire genomes, to trace our past to the origin of our lineage and find our closest ape relative. We now know, of course, that chimpanzees (and their close cousins, the bonobos) are our closest living relatives. Along the way, we'll discover some surprising aspects of our genome that show our deep evolutionary connectedness with all the apes, not just chimpanzees. These studies reveal to us that our genome is like a patchwork quilt, with new pieces added over the course of generations—a genome with segments that were picked up at different stages of our ancestry.

As recently as twenty-five years ago, most anthropologists believed our evolutionary separation from the apes occurred very deeply in the past, all the way back to fifteen million years ago, with the first human ancestor being an extinct robust-jawed and large-molared ape from northern India named *Ramapithecus*. Today, genetic analyses can make far more accurate estimates about the origin of our lineage, pointing to much more recent times for our separation from chimpanzees.

Another long-standing question concerns the evolutionary forces that led to our reproductive isolation from the ancestors we once shared with chimpanzees. Various theories have been put forward as answers, including the traditional and widely believed "savanna hypothesis" initially advanced in 1879 by Charles Darwin in his *Descent of Man* and later promulgated by the paleoanthropologist Raymond Dart, who in 1924 made the first discovery of a human fossil in Africa—the Taung skull—formally named *Australopithecus africanus*. The savanna hypothesis and its variants suggest that a progressive warming and drying of the environment led to an expansion of the African savanna, prompting some early ape species to leave their forested habitats behind. On the savannas, these species developed bipedalism, eventually becoming a new species in the earliest beginnings of the human lineage, while their ancestors remained in the forests. Such geographic separation led to relatively quick reproductive isolation and speciation of early humans from their ancestors. Other hypotheses do not emphasize such clear geographic separation, suggesting that the earliest human ancestors still lived in very nearly the same forested habitats that their ancestors had always lived in and therefore geographic isolation was not the prime mover of the emergence of the human lineage. Researchers have now started comparing genomes of humans and chimpanzees to begin to evaluate the likelihood of these hypotheses and whether this ancient evolutionary split resembled a short and quickie divorce, or more like a long and drawn-out affair with mating between the two emerging species.

Throughout our quest in the last quarter of the twentieth century to determine the precise evolutionary relationship among the apes, a persistent debate—aptly captured by the phrase "molecules versus morphology"—was what evidence was best to use. Would we obtain the most accurate reconstruction of these relationships if we compared many anatomical features of skeletons and skulls among different species, or would comparisons of the biological molecules of DNA provide the most reliable evidence? Since the 1990s, this debate has subsided in favor of DNA, and I will explore the rationale that justifies using DNA and now full genomes to build very accurate reconstructions of species' relationships. Like opening a cloudy and jammed window to let in both light and fresh air, this new perspective has revealed that outward appearances and even the smallest of anatomical details of different species can deceive and even mislead us, and it continues to help us obtain more accurate understandings of the complex processes by which anatomical structures of organisms can evolve.

Ever since Darwin, we have wanted to understand how and why our unique features, which provided distinct advantages on our evolutionary journey since the separation from our common ancestor with chimpanzees, have evolved. Shaped by the mechanism Darwin described as natural selection, these human adaptations increased our ability to survive and enhanced our capability to reproduce and successfully raise our offspring. But what are the adaptations that separate us from chimpanzees, how many of these adaptations do we have, and where in the genome are they found? For example, increased cognition and language almost certainly result from changes that occurred in numerous parts of our genome, but we don't know in how many regions, where they are located, and how these changes led to our ability for complex thought and communication. Beyond our species-wide adaptations, which all humans possess, there are also adaptations unique to peoples living in different parts of the world. These adaptations evolved as modern humans spread out from the site of their evolutionary origin and settled in different regions, encountering different environmental conditions, eating different foods, and facing different pathogens.

Having entire genomes to work with allows researchers to voyage across their vast uncharted nucleotide bases searching for locations that bear signs of having been shaped by the forces of natural selection, and that therefore might represent regions underlying our unique features. Once discovered, such regions become a point of departure for further research into determining exactly how these regions function and if they indeed contributed to our adaptations. After all, for most of the genome, and even for most of our genes, we still have only a relatively simple appreciation for how they influenced our biological features.

DNA analysis is also shedding light on another vexing evolutionary riddle. Ever since the first discoveries of Neandertals in the mid-1800s, there has been a preoccupation with their evolutionary fate. This mystery has grown deeper with time, especially with the introduction of "out of Africa" models of human evolution, from the mid-1980s into the 1990s, which in their strictest form suggested that anatomically modern humans originated in Africa and subsequently spread to Asia and Europe, replacing all archaic forms, including the peculiar Neandertals, and relegating them to the evolutionary dustbin.

In the early 21st century analyses of full nuclear genomes from diverse peoples from around the world overwhelmingly indicate that the evolutionary origin of modern humans was in Africa and that we subsequently dispersed into Europe, Asia, and eventually the Americas. But the theory is now decidedly shorn of the notion that these archaic hominins living in Europe and Asia were completely replaced by newly dispersing modern humans. One of the greatest triumphs in recent anthropological studies has been the recovery of ancient DNA from fossils of extinct relatives. The determination of the full nuclear genomes of Neandertals and of an ancient Denisovan of Asia, a previously unknown contemporary of Neandertals, confirms that modern humans interbred with both of these archaic human relatives. Accumulating genomic evidence presents a fundamental challenge to our previous views that the origin of our modern species in Africa was recent, quick, and regionally restricted.

Many questions still remain about the evolution of modern humans in Africa, however, and the vast size of the genome is providing researchers with ample evidence for rigorous testing of previous hypotheses about human origins. With the help of genomic analysis, we are beginning to have a fairly detailed understanding of our place in the primate evolutionary tree and how other primate species are related to one another. This knowledge provides the essential evolutionary framework by which we can trace our adaptations back in time to learn when on the primate tree they first started to emerge. While certain human features will likely prove to have uniquely evolved along the human lineage, many others presumed to be unique will have deeper origins within our shared past with other primates. We already know that a number of genes that show signatures of adaptation within the time frame of human evolutionary history (for example, several genes associated with our increased brain size) also show signatures of adaptation on the common ancestral stem we share with apes. Charles Darwin himself appreciated this early on, suggesting that "the difference in mind between man and the higher animals, great as it is, certainly is one of degree and not of kind.[3] Results from the analyses of the genomes of many other nonhuman primates, and even distantly related animals, will likely be a lesson in humility, revealing our deep connection to the rest of the animal world.

My aim in writing this book is not to provide definitive answers to the persistent questions we have about our evolutionary origins—we can only hope to approximate answers to these questions more accurately—but to introduce the reader to the ways in which genomes can be used to begin to reexamine old questions with new evidence. Our evolutionary past is a puzzle with disparate and fragmentary pieces remaining—including fossils, cultural artifacts, and now genomes—that ultimately all need to fit together to give us a coherent and internally consistent view of our history. As we shall see, genomes are providing us with a powerful new tool that we can use to bring the puzzle much more clearly into view.

Ancestors in Our Genome

CHAPTER 1

Looks Can Be Deceiving

When I was in graduate school in physical anthropology in the 1990s, there was a war around me—well, let's say a battle—being waged between those of us who studied the anatomy of bones and fossils and those who studied genetics. We even occupied different floors in our department, which only further heightened the divide. So who had the high ground? Well, from my perspective, it was the ninth-floor morphologists like myself who were studying anatomy. By studying the anatomy of fossils from extinct primates, some of them our close relatives, we could determine if they ate fruits, leaves, or grass seeds. We could tell if they climbed in trees or walked on the ground, how much they weighed, and how big their brains were. We also felt we had the upper hand in determining the evolutionary relationships of primates—the details of the skeleton could tell us more about the kinship among different species than any single gene. Plus, on the ninth floor we had nice sunny views of lower Manhattan, which added a little to our sense of self-importance!

Genetics, on the other hand, could tell us nothing about how a primate moved around in its habitat, how much it weighed, or what it ate. So what was it good for? What could it reveal about our evolutionary history? The debate between morphology and genetics played out on a larger stage at the annual meetings of the American Association of Physical Anthropologists and in our professional journals. The main contentious issue centered on whether genetics or morphology was better for determining the evolutionary relationships of the primates. A prominent book published at that time, *Molecules versus Morphology: Conflict or Compromise?* by Colin Patterson,[1] polarized the debate and helped fuel the arguments.

For relief from the battle, I decided to go fishing. Not fishing in the literal sense, though I did escape from reading and measuring bones occasionally to visit the shore. I went fishing for a dissertation project. My main interests at the time were in studying the morphological diversity of primates and humans to reconstruct how extinct primates moved, using evidence gleaned from their fossilized remains. I became especially interested in studying anatomy in order to reveal the evolutionary relationships among one group of our African relatives, the papionin monkeys, which includes the long-faced and terrestrial baboons, mandrills, and their relatives. An accurate evolutionary tree for this group was in doubt because morphological evidence was pointing in one direction and some new genetic evidence was pointing in another. Since bones and muscles were all I knew at the time, I planned to do a thorough analysis of the anatomy of this group of monkeys. The way in which my research played out, and my personal ideological evolution during that time, revealed several very important lessons directly applicable to our understanding of the evolution of great apes and humans.

ENTER THE MONKEYS

Monkeys and apes from Africa and Asia are grouped together due to the fact that they share derived and newly evolved details of their ear anatomy, the evolutionary loss of a premolar tooth as well as many genetic aspects. But Old World monkeys and apes separated from one another about twenty-four million years ago. The papionin monkeys are a subgroup of these monkeys that began to differentiate into different species at about the same time that the hominoid apes— chimpanzees, gorillas, and humans—were splitting into different species. No doubt the two groups of primates came into contact on the African plains, and very likely sometimes on less than friendly terms. There is archeological evidence from the site of Olorgesailie in Kenya that our human ancestors, most likely *Homo erectus*, butchered and ate (somewhere in the time frame of 700,000 and 400,000 years ago) a now extinct giant relative of the living papionin monkey called the gelada.[2]

The African papionin monkeys include three large-bodied, long-faced, and very terrestrial members (the geladas, mandrills, and baboons). It also includes two smaller-bodied and shorter-faced monkeys, known as the mangabeys, which are partially terrestrial to arboreal in their habits. Up until the early 1990s, morphological studies had concluded that the larger-bodied and long-faced monkeys were the closest evolutionary

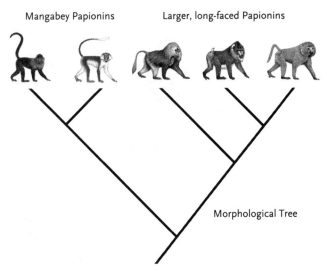

Mangabey Papionins   Larger, long-faced Papionins

Morphological Tree

Figure 1.1: Until the 1990s, most anthropologists believed the papionin African monkeys were evolutionarily related to each other in the manner depicted by this species tree.

relatives, whereas the two smaller and shorter-faced mangabeys formed a separate group (Figure 1.1).

As I was fishing for my subject, I decided to study the papionins, in part, to resolve a lingering riddle about their relationships. During the 1970s, scientists in London studying blood proteins had found evidence that indicated a close similarity between one of the mangabey groups and baboons, which essentially would disrupt the close relationship that had been established for the long-faced monkeys. Clinging to my morphology perspective, I thought this confusing result was due to the examination of just one protein out of many many thousands, so I decided to map out a full-fledged morphology project in which I would reexamine the details of the entire skeletons (from the skull down to the tail) of all the papionin monkeys. Surely, the answer was in the bones, if enough anatomical detail could be analyzed!

At the same time, my graduate department at New York University had just hired a new professor, Todd Disotell, who had recently published his dissertation on the evolutionary relationships of papionins determined through a genetic analysis. In his research, Disotell had collected and analyzed DNA sequences from a gene from the mitochondrial genome, the small circular genome contained within the cell's mitochondrion (Figure 1.2). The genome is inherited only through maternal lineages—mothers to daughters to granddaughters and so on. DNA studies in the 1990s mostly focused on mitochondrial genes, since they were relatively straightforward to isolate in the laboratory and we knew that they had

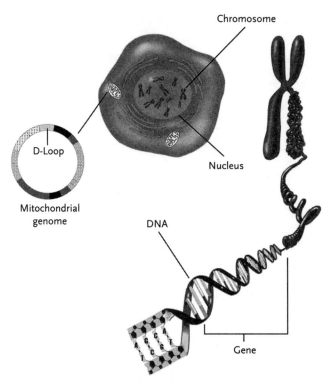

Figure 1.2: A cell showing pullouts of the nuclear genome and the mitochondrial genome.

undergone mutational change in the past at a faster rate, a useful feature if one wishes to detect DNA differences between species. One of the tantalizing aspects of Disotell's study was that it pointed to the same close evolutionary relationship between one of the mangabey groups and the baboons as was found in the 1978 blood protein study, and he proposed a new evolutionary tree for these monkeys (Figure 1.3).[3]

To me, this new finding seemed to make it even more pressing to examine the papionins in greater detail, but from the perspective of the bones, of course. However, as I pored over the anatomical literature for papionins, which extended back about one hundred years or more, I started to realize that there was probably little opportunity to find something new that would add significantly to what we already knew about papionin morphology and their evolutionary relationships. Perhaps I could study the postcranial skeletons (the skeleton from the neck down) in more detail, but the details of the skull and facial skeleton (the sizes and shapes of the upper and lower jaws, the bones of the nasal openings and the eye sockets) of these monkeys had been analyzed already in a number of evolutionary studies. In any case, I started to believe that my morphological

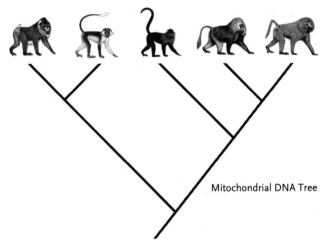

Figure 1.3: Mitochondrial DNA was among the first evidence to point to a breakup of the close relationship among the three large-bodied and long-faced papionin monkeys, and also a breakup of the close relationship between the two smaller-bodied and short-faced papionin monkeys, known as mangabeys.

analysis, no matter how detailed, would only support the evolutionary relationships that morphologists had already established, although I continued to believe that it was through morphology and not molecules that we would continue to gain greater understanding of evolutionary relationships among different species.

HOMOLOGY AND FAMILY TREES, PART ONE

To understand how we uncover evolutionary relationships among species—their family trees—we need to look at the concept of a *homologous* feature. A homologous feature is a feature shared between two or more species because they inherited this feature from a common ancestor. Examples of homologous features are the bones that make up the arm, forearm, and hand of all mammals. These animals all share the same set and arrangement of bones: an upper arm bone (humerus) closest to the body, two forearm bones (the ulna and radius) just after the arm, and five finger bones at the very end of the forelimb. In some mammals, like whales and bats, some of these bones have been evolutionarily reshaped, elongated, or diminished through the process of adapting to their different life ways, but the basic set and arrangement is still the same. All mammals share this basic pattern of bones in the upper limb because they inherited it from their common ancestor.

While the forelimb bones of all mammals are homologous features, these features are not newly evolved just in the class of mammals. In fact, all tetrapods—the four-limbed amphibians, reptiles, birds as well as mammals—have the same set and arrangement of bones. In mammals, we call these features primitive homologous features since they evolved in the distant common ancestor mammals share with the other tetrapods.

One way that we build evolutionary trees is through a method called cladistics, which organizes organisms into groups, known as clades, containing all organisms that have descended evolutionarily from the same common ancestor. The mammals form a clade that includes the placental mammals such as primates, bats, whales and dolphins, seals, giraffes, elephants, sloths, anteaters, rodents, and so on, as well as diverse marsupials and the egg-laying mammals (the duckbilled platypus and echidnas). Despite their amazing diversity, all mammals form a single clade because they share the same features inherited from their common ancestor: hair or fur, endothermy (their bodies produce heat), three middle ear bones (the malleus, incus, and stapes), and the presence of mammary glands in females. Thus clades are defined by special homologous features newly evolved in the last common ancestor of the group and inherited by all members of the group.

For example, the presence of a tail is a primitive homologous feature in all primates, inherited from the remote common ancestor they share with fish. Reptiles and amphibians also inherited a tail from fish, even though some, like the frog, have lost it. While most primates possess tails (even if they are very short in some species), all apes and humans have lost an external tail. Thus, the loss of a tail is a shared derived feature within the primate group that is cladistic evidence of the close relationship among all apes and humans, to the exclusion of other primates. While the loss of a tail in all apes and humans is evidence they are all united within a common clade, the feature does not tell us anything about the relationships among the different members of this group. A scientist would have to conduct a more detailed study of the features of each member of the group to resolve these relationships.

There are other features that look similar between two organisms, not because they were inherited from a last common ancestor, but because they evolved independently to have similar functions. These features are called analogous features and arise by the process known as homoplasy. Popular examples of such features are the wings of bats and of birds. The wings of these two different animals are not homologous structures, because these two organisms did not inherit their wings from a common ancestor. Bats are milk-producing, endothermic animals of the class *Mammalia* and birds are egg-laying and feathered animals of the class *Aves*. The wings of these unrelated organisms are anatomically very different, and evolved independently in the organisms because they provided a beneficial adaptation. When homoplastic features occur in species that are distantly related—like bats and birds—they are called evolutionary convergences. Another example of evolutionary convergence is the evolution of a fusiform body shape and fins (or flippers) in fish and the distantly-related mammals such as dolphins and whales. Analogous or homoplastic features can also be found between much more closely related species, and are called parallel features. The detailed resemblance between the bones of the shoulder joint in some large-bodied New World monkeys like spider monkeys and the bones of the shoulder joint in apes is one well-known example.[4] In both cases, the joint structure gives them a great deal of shoulder mobility and allows them to hang by their forelimbs beneath branches. However, it should be realized that sometimes parallelisms are a result of common growth tendencies inherited from a common ancestor (more on this later). The problem with using homoplastic features to ascertain phylogenetic or evolutionary relationships, especially those between closely related species, is that they can be misinterpreted to suggest a close evolutionary relationship even if there is none. Based exclusively on an examination of the shoulder anatomy of a spider monkey (a New World primate) and the shoulder of a gibbon (an Asian ape), for example, we might conclude that they were closely related, when in fact they are not.

HOMOLOGY AND FAMILY TREES, PART TWO

If homology is thus a less than straightforward indicator of relationships among species for morphology, is homology any clearer for DNA evidence? Since homologous features are by definition those features in a set of species that have been inherited from a common ancestor, it may be helpful to look at exactly what happens during biological inheritance from parents to their offspring. We might say of a child, "Look, he's got his father's ears!" However, what the child has really inherited from a parent is not such outwardly observable features as eye color, ear shape, hand shape, and so on, but rather the unique package of inherited instructions that serves as the blueprint for the development of such features. When we talk about a homology between two species, we are actually referring to the DNA (or information) they each possess today that was inherited over generations since their most recent shared ancestor. To ascertain homology (particularly a shared derived homology that can be used to determine if two species are most closely related), we need to compare what is actually being inherited—their DNA. Is it any easier to identify homologous and homoplastic features in DNA sequences, compared to morphological features? Under certain circumstances we can say yes, especially for closely related species for which we can assume that only minimal DNA differences occurred between them since their last common ancestor. This is true because the DNA nucleotides we study are the same in whatever species we are comparing. In other words, there are not different forms of adenine (A), and the same goes for cytosine (C), guanine (G), and thymine (T). Whether in a papionin monkey, a great ape, or a human, an adenine molecule is the same molecule in each species. A morphological feature that looks very similar in two different species, on the other hand, may in fact have a different genetic basis in the two and therefore a different evolutionary origin. The field of genetics has not, by and large, been able to determine the exact genetic basis for most morphological features. When we are comparing DNA nucleotides we know when we are comparing apples with apples or oranges with oranges. With morphological features, one could very well be comparing apples with oranges.

Examples from recent investigations show us how the genetic basis underlying very similar-appearing anatomical features may be due to very different genetic changes in each of them. One recent example comes from genetic studies of Atlantic beach mice from Florida by Hopi Hoekstra at Harvard University.[5] Populations of these mice on the mainland of Florida are dark brown in coloration (similar to soil colors), but the beach mice populations inhabiting the white sand dune islands of the Atlantic and Gulf Coasts of Florida have each evolved very light-colored coats, providing them with camouflage against the sandy background. A simple expectation would be that the genetic changes underlying the light coloration in both populations would be the same. But surprisingly, when Hoekstra searched for the genetic change identified as causing light coloration in the Gulf Coast beach mice, she could not find it in beach mice from the Atlantic Coast. In fact, she discovered that the genetic basis of the light-colored coats in mice from the different coasts must lie in separate genes altogether. There are other examples too. Pelvic reduction evolved multiple times in different populations of stickleback fish, though it is due to different mutations (albeit within the same gene, Pitx1).[6] Complete loss of pigmentation is found in separate Mexican cavefish populations, although the depigmentation has different specific underlying genetic causes.[7] Thus, there can be quite different genetic ways of producing very similar morphological features[8]—not a good thing if you want to determine species' evolutionary relationships on the basis of anatomy.

Another advantage of molecular data is that it requires fewer judgments beforehand about which DNA nucleotides are homologous. Almost all morphological studies make judgments of homologous features among different species before the evolutionary analysis begins, and this occurs no matter how carefully morphologists try to avoid bias in their analysis. Although issues are not so straightforward, analyses can become circular, because the morphologist is already influencing the results, basically deciding ahead of time which features are homologous and which features are not. But if we are certain we have determined the DNA sequence in the same gene in all the species in our analysis, then judgments about which DNA nucleotides are homologous and which are homoplastic are decided on the basis of pure calculation. The methods we use to compare DNA regions from different species evaluate all possible gene trees for a given number of species and find the single tree (or several trees) that requires the least number of past DNA changes or mutations. So for example, if we are studying the same gene from three different species, there are only three different possible ways the three genes could have branched from one another. We then follow a method based on the principle known in science as "Occam's razor," or parsimony—that the simplest explanation is usually the correct one—and assume that the tree requiring the least number of DNA changes is the tree most likely to be correct. After the best tree is found, any DNA changes that do not support the tree are usually interpreted to be homoplastic. Thus, the method of finding trees using the principle of parsimony sorts past DNA changes into those that are likely to be homologous and those that are likely homoplastic, making the procedure arguably more objective than determining trees using morphological features.

There's another problem for morphologists, which concerns how to determine which features to study to determine evolutionary relationships. Features that appear to be separate may in fact be functionally, structurally, or developmentally interrelated. For example, a large-diameter femur (thigh bone) is also likely to have a large femoral head (the bony ball that fits in the hip socket), since both features are associated with supporting body weight. In a facial skeleton that is very long, as it is in baboons and mandrills with their drawn-out muzzles, it is usually the case that the nasal opening in the facial bones is long and narrow. We wouldn't want to count both features—long face and long and narrow nasal opening—as separate pieces of evidence in a morphological analysis to indicate a close relationship between baboons and mandrills. The two features are not independent of one another. With molecular data, it may be easier to avoid this problem of functional interdependence.

There is a further advantage to using DNA to establish species relationships. Comparing different segments of the genome, such as DNA from different chromosomes, or different parts of the same chromosome, allows us to build numerous, separate gene trees for species, thus testing their relationships using truly independent pieces of genetic evidence. Think of this advantage like a police detective might. Can you make a good case against a suspect when all you have is a single piece of evidence? Maybe, but you can make a much stronger case when you have multiple pieces of evidence from different independent sources, all pointing towards the same suspect. The more independent genetic evidence—or separate DNA gene trees—that point to the same relationships among a series of species, the more credible it becomes that we have indeed determined the actual species relationships. Moreover, since there are literally hundreds of thousands of these different independent segments in the genome, it's clear that the genome beats morphological studies for evaluating species relationships because of the sheer magnitude of the evidence.

BACK TO THE MONKEYS

Faced with all these arguments for using DNA evidence to build species trees, I began to reconsider my plans for a morphological study of the papionins. The idea of congruence between independent pieces of genetic evidence appealed to me as an overwhelming way to establish the true evolutionary relationships of these monkeys. After all, perhaps the mitochondrial DNA tree was a fluke? So Disotell and I started to put together a scheme to test his mitochondrial DNA tree by involving more DNA sequences from these monkeys, but from genes located on different chromosomes in the much larger genome.

After a few years of lab work—work that today could be done in several weeks!—the five separate gene trees we identified turned out to be congruent with each other. Furthermore, the species relationships they supported were the same as those supported by the mitochondrial DNA results. The large-bodied and terrestrial long-faced baboons and mandrills were split up into different lineages. So were the much shorter-faced mangabeys, with one group being more closely related to baboons and the other group most closely related to mandrills. The statistical reliability of Disotell's original tree had been corroborated through the consilience of various pieces of independent genetic evidence.

Now that we had a very reliable species tree built from DNA data, what were we to do about the morphological features of the papionins that did not "agree" with this tree? One day I was chatting about this problem with another professor at NYU, Clifford Jolly, on a tea break on the fourth floor (the "genetic zone" of our department). Cliff is one of the most erudite of biological anthropologists I have known and, lucky for me, he was one of my graduate mentors. That day, he mentioned to me (here I paraphrase)— in his understated yet amusing and insightful way—that "indeed, one can hang morphological traits onto the molecular tree as if hanging ornaments on a Christmas tree." In other words, we needed to take a good look at the morphological features that seemed to support the morphological tree, and reexamine them in light of the DNA tree.

Fortunately, many morphologists embraced Jolly's adage and several studies immediately followed, in which large data sets of morphological features from primates were used to build evolutionary trees. In one study in 2000, by anatomists Bernard Wood at George Washington University and Mark Collard at Simon Fraser University, features from four different regions of the skull were examined—from the upper jaw and roof of the mouth, the lower jaw, the bony parts of our face, and the top and the bottom of the braincase—so that separate morphological trees could be built for features from each region that could then be compared to the molecular tree.[9,10] In a theoretical shift, the study accepted *a priori* that the molecular tree was correct. Unexpectedly, features from *all* regions of the skull supported morphological trees that were different from the molecular tree. Wood and Collard were insistent in pointing out the meaning of all this: that features of the skull, jaws, and teeth can totally mislead us as to the true evolutionary relationships of monkey and ape species. This was a shocking revelation to many paleoanthropologists, which would have gone unrevealed if not for the DNA analyses.

The seeming incongruence between papionin monkey morphology and DNA results led morphologists to more closely examine anatomical features in the papionins. Several years after the results from the DNA studies conducted by Disotell and myself were in, anatomists John Fleagle at Stony Brook University and his then doctoral student Scott McGraw found a limited number of detailed anatomical features of papionin monkeys, from the arm and leg bones and dentition, that agreed with the species relationships based on DNA.[11] However, without the prior DNA results, it is difficult to say whether these details of the skeleton would have been interpreted to support the relationships discovered using DNA methods. Indeed, Fleagle and McGraw indicated the difficulty of the problem when they described their results as "unmasking a cryptic clade," meaning papionin evolutionary relationships are difficult to detect on the basis of just anatomical evidence. And more recently, Chris Gilbert at Hunter College/City University of New York and his colleagues carried out some very sophisticated morphological analyses to isolate the drastic effects that differences in body size can have on altering the shapes of anatomical features, confirming that these morphological features also agree with the DNA results.[12] Being able to remove the effects of body size differences is extremely important when comparing primates, since the larger papionin species (the baboons, mandrills, and geladas) are more than twice the body weight of the smaller monkey species (the two mangabeys).

What can these results from morphological analyses tell us about interpreting species relationships using fossil evidence? Whether we're studying monkeys, apes, or human ancestors, fossil remains consist of bone fragments from different regions of the skull, parts of the jaw, and different parts of the skeleton. Paleoanthropologists carefully reconstruct these fragments and painstakingly analyze their details. Only by studying fossils can we infer the size and shape of ancient members of our family tree and learn about what foods they ate and their styles of locomotion. But often these fossils are only partial or distorted and from a single or a very small number of individuals. If we cannot reliably tell the correct evolutionary relationships of even living primates, for which entire skulls, dentition, and skeletons of each species can be measured, and for which the skeletons of many different individuals can be studied, then how can we expect to be able to determine the correct evolutionary relationships of our primate and human fossil ancestors?

One of the most significant debates in paleoanthropology centers on interpretations and counterinterpretations of where exactly these fossils fall on our evolutionary tree. We want to know whether a particular species is a direct human ancestor, or is on a side branch, and how any particular species is related to other human-like species. And while we are confident that we have identified many relatives on the human family tree, for even the most heralded human fossils from species like *Homo habilis* ("Handy Man"), *Australopithecus afarensis* (known by the famous "Lucy" skeleton), *Paranthropus boisei* ("Nutcracker Man"), and others, our precise understanding of their relationships with other early human species may be continuously stymied by misleading and hard to interpret morphological features. Homoplasy among our fossil ancestors has made it difficult to work out the specifics of who is related to whom.[13,14] To unravel our evolutionary relationships among the living apes, we turn now to ape and human DNA evidence.

CHAPTER 2

Many Trees in the Forest

The DNA Quest to Find Our Closest Ape Relative

Through the last quarter of the past century, anthropologists were struggling to answer a central puzzle—who was our closest living relative among the hominoid species? Hominoid is short for *Hominoidea*, the superfamily designation for a group of primates that includes humans, the three great apes—Asian orangutans, African gorillas, and chimpanzees (and their very close cousins, the bonobos)—as well as the lesser apes, the siamangs and gibbons of Southeast Asia. If we could know the order in which each species branched from one another we could better understand how their anatomical features evolved. For example, if it turned out humans are more closely related to chimpanzees, this could mean that we too were once knuckle-walkers (like the two African great apes are) before we became bipeds. However, if chimpanzees and gorillas are more closely related to each other, than there is no reason to suspect we passed through a knuckle-walking phase. Fossil-hunters crave this information so that they know what traits to look for in the earliest human fossils.

By the early 1960s, on the basis of immunological experiments and protein comparisons between primates, it was already increasingly clear that orangutans were more distantly related to the African apes and humans, and the lesser apes more distantly still.[1,2] The questions that continued to irk anthropologists were which of the two African apes—the chimpanzee or gorilla—was most closely related to humans, or were humans on a branch well separated from the two species of ape? Because we could not determine the order in which the species branched apart from one another, their relationship was represented by a three-way split, the "hominoid trichotomy," on our evolutionary tree (Figure 2.1). For many years, we were stuck with this tree, where all three species were equally closely related to each other, although evolutionary theory suggested that new species evolve through a two-way (dichotomous) splitting process. The "hominoid trichotomy" problem was thought to be so intractable, that some major textbooks suggested, until quite recently, that the exact relationships among these species might never be understood.

The problem of the relationships of the great apes to humans runs deep. Historically, some researchers wished to prove that humans were separate from, or perhaps "better" than, the apes, pointing out our very distinct features. Even today many authors of textbooks still use a taxonomy—or system of categorizing species by name—that places all the great apes (orangutans too) within a group separate from the group for humans. This is mainly due to their desire to recognize that the great apes share a very similar "grade" of evolution that contrasts starkly with the human condition. Apes are all largely forest-living, possess strong arboreal adaptations (long and strong arms and hands for tree climbing), are very hairy, have similar-sized brains, and are quadrupedal when on the ground. But I believe these taxonomies can also be attributed to a lingering anthropocentric recalcitrance to admit our close evolutionary relationship to the apes. Darwin did not have this problem, however. In 1871 in *The Descent*

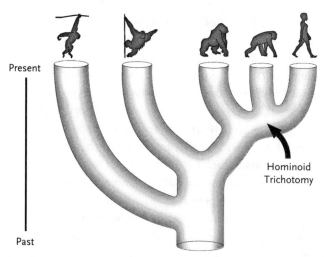

Present

Past

Hominoid
Trichotomy

Figure 2.1: Evolutionary tree of the apes and humans. Up to the late 1990s, scientists were confused about the evolutionary relationships among humans, chimpanzees, and gorillas and wondered which two of these primates were most closely related. The "problem" was represented by a three-way split among them and was referred to as the "hominoid trichotomy."

of Man,[3] he wrote, "these two species [gorillas and chimpanzees] are now man's nearest allies" (p.161).

Despite the results of early protein and immunological studies, some biologists (mostly trained in anatomy) persisted in suggesting that all three great apes—orangutans, chimpanzees, and gorillas—were evolutionarily more closely related to each other than any one of these apes were to humans. For example, in the early 1980s, one prominent zoologist, Arnold Kluge at the University of Michigan, confirmed this view by basing his analyses on the "total" evidence of hominoid skeletal features and soft-tissue structures of the skin and reproductive system. His study excluded any genetic data, however—so much for "total" evidence!

STICKY GENOMES AND A STICKY PROBLEM

Throughout the 1970s and 1980s, the field of DNA research continued to grow at a ready clip. The late Charles Sibley and his then student Jon Alquist, at once both ornithologists and molecular biologists, had already invented a major new technique called DNA-DNA hybridization, in which entire genomes from different species are experimentally "glued" together or "hybridized" in the laboratory, and examined for the degree to which they stick together. This procedure relies on the fact that DNA has two complementary strands that can be "unzipped." Unzipped strands from one or more species can be made to "rezip" to each other; the more closely they bind, the more closely related the species.

Sibley and Alquist first applied their DNA-DNA hybridization to a vast array of different bird species, and their findings ruffled the feathers of fellow ornithologists when several strongly held ideas about how certain birds were related to each other were not supported. Since the researchers believed their technique was something of a panacea for determining taxonomies generally, they soon moved beyond birds to apply their technique to other animals, including humans and apes. The results from DNA-DNA hybridization in hominoids found that chimpanzee and human genomes stuck together more strongly than either the human-gorilla genomes or the gorilla-chimpanzee genomes.[4] However, this conclusion raised the hackles of more than one anthropologist. Some had some legitimate concerns about how the hybridization experiments were carried out; others felt the results conflicted strongly with the series of anatomical features—such as thin tooth enamel and multiple bony features of arms, forearms, wrists, and fingers, all related to a knuckle-walking style of locomotion—that seemed to unite the African apes on a branch separate from the human branch.

The debate surrounding the technique of DNA-DNA hybridization reflected a long-standing disagreement about how to reconstruct evolutionary relationships. One school, called phenetics, relied on the degree of certain general similarities between species to indicate their relative evolutionary closeness and group organisms into taxonomic categories such as species, genus, family, order etc. Phenetics is a method that stretches all the way back to the 1700s and to Carolus Linnaeus, the father of taxonomy. In earlier anatomical studies, phenetics made measures of whether species resembled one another in such features as the overall proportions of the skull and skeleton, overall body proportions, and overall outward appearance. In early DNA studies, phenetics meant comparisons between the overall size and electric charge of proteins, the immunological interactions between their proteins or, as in DNA-DNA hybridization, the strength with which two species' genomes stick together.

The other school of evolutionary reconstruction relied on cladistics, mentioned previously. Cladistics is a more recent methodology developed in the 1970s by the German entomologist Willi Hennig. This method proposes that only *some* similarities between two organisms are of significance in indicating that they are each other's closest evolutionary cousins, or in Hennig's words *Schwestergruppen* ("sister groups"). These similarities, called shared derived features, are features that were newly evolved (derived) in the immediate common ancestor of two species and were inherited together (shared) by these two sister species.

Importantly the shared derived feature must be a modification of the ancestral condition of this feature. For example, all five hominoid groups (seen in the illustration above) lack an external tail, distinguishing them from the over two hundred other primate species in the world who sport such tails. It's likely, then, that the evolutionary loss of an external tail must have been relatively recent in the evolution of primates; specifically it must have occurred in the last ancestor shared by all the living apes. The loss of a tail is an informative cladistic feature that indicates they are all more closely related to each other than to any other primate species. On the other hand, all hominoids have two premolars (bicuspid teeth), having lost one of the three premolars that the earliest primate ancestor had and which most primitive primates still retain. Yet Old World monkeys also have only two premolars, suggesting a premolar was lost in the common ancestor of both of these higher primate groups, Old World Monkeys and hominoids. Following cladistics then, the presence of only two premolars in hominoids is not unique to them and therefore could not be used as evidence to support their close relationship within a clade.

Cladistics did not really gain a foothold in anthropology until the mid- to late 1980s, but its rise in prominence had a lot to do with why DNA-DNA hybridization results were questioned. The "stickiness" of genomes from two species was a phenetic measure; it did not distinguish shared derived features (the cladistically meaningful kind) from more primitive shared features that are cladistically meaningless in terms of determining closeness of evolutionary relationships between species. There were other problems too. The DNA hybridization method seemed better at determining the relationships between species that had branched away from each other in the more distant evolutionary past, and not for those species that had branched from each other more recently—like humans, chimpanzees, and gorillas. Moreover, some reinterpretations of the hydridization data using slightly different methods and statistical tests indicated that the different branching arrangements among the three apes were essentially indistinguishable. The problem of the hominoid trichotomy continued to raise its ugly head.

BUILDING GENE TREES

The new molecular findings that began to trickle into laboratories in the 1990s, including my own, were actual DNA sequence data, down to the nitty-gritty repeating nucleotide bases—the A, C, G, and Ts—that make up the rungs of the ladder of DNA. Once the unique order of these bases is known for a short region of the genome, we refer to it as a DNA sequence. It turns out that cladistic methodology is perfect for analyzing DNA sequences, because it fits nicely with the way DNA sequences evolve over time and the way we believe that new species originate. For example, DNA often evolves through mutations that change a particular DNA base to a different base—for example, an A to a T, or a G to a C, although any mutational change among the four bases is possible. When these mutations occur between species—a chimpanzee may have an A in the particular position in the genome where a human might have a T—they are known as base substitutions. Sometimes the mutation involves the deletion or insertion of a base or more between species. These base substitutions, deletions, and insertions are the "features" used in cladistic analyses of DNA.

Some of the earliest DNA sequences compared between hominoids were from the gene called *alpha 1,3 GT*.[5] This gene codes for an enzyme that helps produce a protein found in the cell membranes of mammals but for some unknown reason the gene was inactivated in monkeys and apes. Since inactivated genes have lost their function, they accumulate mutations more quickly than still active genes, which will produce more DNA differences between species at this gene. For the majority of the gene's 371 bases, the species were identical. But 21 of the bases (see Figure 2.2) show differences between species.

Molecular theory suggests that any DNA differences between the species that we see today must have arisen through the process of mutation in the past, as the species branched from each other. Like breadcrumbs left behind to mark one's trail through a forest, these mutations can be used to reconstruct the order in which the species separated from each other in the course of evolution.

Since the earliest species among this group to branch was the orangutan, in the analysis of *alpha 1,3 GT* gene, the orangutan (known by the scientific name *Pongo*) is designated as the out-group species. The hominoids, represented by their scientific names *Pan* (chimpanzees), *Homo* (humans), and *Gorilla* (that's an easy one), constituted the in-group of special interest (Figure 2.2). As expected, there were many positions along the gene where all the in-group species shared the same base while the orangutan gene contained a different base. This pattern indicates that all in-group species evolved this particular base in their shared common ancestor, and points to the fact that they are all more closely related to each other than any is to the orangutan. Unfortunately, these sites do not tell us anything about the relationships among the in-group taxa, our main quest. For this, it is necessary to look at twelve sites that vary within the in-group.

These twelve sites (asterisked in the illustration) can be divided into two categories: sites that are informative about evolutionary relationships and sites that are not. Uninformative sites occur only in a single genetic sequence. The T in *Gorilla* would be one example of one such site. Because the substitutions, insertions, or deletions that appear at these sites are not shared with another sequence, they cannot tell us anything about the relationships of these sequences. This leaves us with only two sites, numbers 186 and 229. For these two sites, the bonobo and human sequence share the same base or have deleted a base, while the gorilla and orangutan share a different base. The sites can then be mapped onto three different trees that represent the possible relationships among these species. Using the principle of parsimony, we can evaluate the three trees to find the tree with the least number of mutations necessary to explain the pattern of shared DNA bases.

The simplest tree (requiring only two mutations) is tree number three in which the genetic sequences from the bonobo and human were more

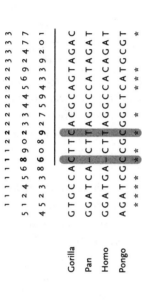

Alpha 1,3 GT Gene

```
        1 1 1 1 1 1 2 2 2 2 2 2 2 2 2 3 3 3 3
        5 1 2 4 5 6 8 9 0 2 3 3 4 4 5 6 9 2 4 7 7
        4 5 2 3 3 8 6 0 8 9 2 7 5 9 4 3 3 9 2 0 1

Gorilla G T G C C A C T T C A C G C C A G T A G A C
Pan     G G A T C A – C T T A G G C C A T A G A T
Homo    G G A T G A – C T T A G G C C A C A G A T
Pongo   A G A T C G C G C G G G C T C A T G C G T
        * * * *     * *       *     * * *   *
```

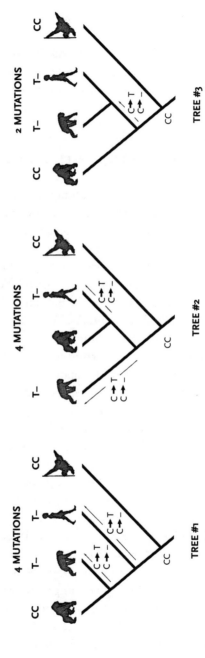

Figure 2.2: Above, the DNA sequences for the *alpha 1,3 GT* gene from the three great apes compared with this gene in humans. Only twenty-one DNA sites of the gene are shown because it is only at these sites that the DNA sequences differ. The two sites that link humans and chimpanzees together are shown in bold font. Below, the three alternative relationships for humans, chimpanzees and gorillas. Tree 3 is best supported because it requires the fewest DNA changes (only two changes compared with four changes in Trees 1 and 2).

closely related to each other than to the gorilla sequence. Under this potential evolutionary history, humans and bonobos must have inherited the deletion of the base at site 186 as well as the T at site 229 from their recent shared ancestor, after the gorilla sequence already branched from the great ape tree. Examining the two other possible trees, we see that each of them requires both the deletion and the substitution to occur in two different places on the tree, bringing the total number of mutations on the tree to four. It's a less parsimonious view of hominoid history, which we think makes it less likely to have occurred.

THE TRICHOTOMY PROBLEM SOLVED?

Well, the hominoid trichotomy "problem" was not to be solved so quickly. Indeed, there were already conflicting results. In 1989, two years before the *alpha 1,3 GT* results, DNA sequences had been determined for the involucrin gene, which codes for a protein in skin.[6] The results showed support for a hominoid tree (Tree 1 in Figure 2.2) in which chimpanzees and gorillas were more closely related to each other than to humans. There were also conflicting results from the immunoglobulin C *alpha 1* gene, which codes for an antibody protein. This gene supported Tree 2, showing gorillas to be the closest relative to humans.[7]

In 1991, however, further support for an exclusively human-chimpanzee clade came from a study conducted by a team of biological anthropologists at Harvard University, led by Maryellen Ruvolo.[8] While previous DNA sequence studies had focused on genes contained within the nuclei of cells, these researchers determined the DNA sequence of the *COII* gene from the very small mitochondrial genome. Many mitochondrial genes are known to mutate faster than genes from the nuclear genome and seemed ideal to address the recent branching in the hominoid family tree. Indeed when Ruvolo's mitochondrial sequences were analyzed, the tree pairing chimpanzees with humans was again found to be the most parsimonious.

In 1992 another study emerged. It focused on DNA sequences generated by the late Morris Goodman, the father of molecular anthropology, and his group of researchers at Wayne State Medical School. One of Goodman's aims during his long and important career was to fulfill Darwin's dream to understand the tree of life for all living organisms. Goodman's group determined the DNA sequence for the beta globin cluster of genes, a set of six genes positioned very close together on Chromosome 11, that code for hemoglobin proteins inside our red blood cells.[9] The length of the globin DNA sequence was much longer than that determined for any previous gene—about forty times as long—and Goodman and his students sequenced it in a human, a gorilla, a chimpanzee, and bonobo, as well as an orangutan. When they analyzed the sequences using similar methods as for the *alpha 1,3 GT* and *COII* genes, the most parismonious tree again supported the closest relationship as between chimpanzees and humans. Now, with *alpha 1,3 GT, COII,* and the beta globin cluster (and several other genes not mentioned here) all pointing to Tree 3, it began to look like resolution might be within sight.

But if there is one powerful thing that motivates scientists, it is the surprises that often lurk in new results. In 1993, the two protamine genes that code for nuclear proteins were sequenced for the hominoids.[10,11] Each gene was claimed to support two different hominoid trees—protamine 1 supported a close relationship between humans and gorillas (Tree 2) and protamine 2 supported a close relationship between chimpanzees and gorillas (Tree 1). Instead of a resolution to the hominoid trichotomy, an apparent conflict among different genes was becoming apparent, raising some thorny questions. How could analyses of the actual DNA bases of genes, the simplest and precise units of DNA, yield conflicting results? Do some genes not "tell the truth"? How could the evolutionary trees built from different genes be absolutely in conflict with each other?

Thus far, we've relied on phylogenetic methods to build evolutionary trees of species. To fully determine a gene tree, however, scientists have turned to another field, called population genetics, to understand the processes by which genes are transmitted from one generation to another, and to learn which evolutionary processes have led populations to separate over time to form new species. To population geneticists, species contain many members possessing genes that differ in subtle ways from each other, but which are all capable of interbreeding and producing viable offspring. Population geneticists don't focus on a single gene or even several genes, but aim to understand all the genes in the genome and in many individuals. They realize that only through the study of many genes can we really get to know the processes by which we evolved.

It turned out that anthropologists who were interested in phylogenetic questions such as the hominoid trichotomy would have a lot to learn from population geneticists. Indeed population geneticists Pekka Pamilo at the University of Helsinki and Masatoshi Nei at Pennsylvania State University had in 1988 developed a theory that could explain why different genes within our genomes can sometimes have completely different evolutionary histories from each other, and why some gene trees will sometimes differ from the evolutionary history of the species carrying these genes.[12] This is known as "the gene tree-species tree problem." Gene tree-species tree incongruence can affect any group of species. Besides hominoids, it has been a particularly thorny problem in trying to unravel the evolutionary relationships among such animals as cichlid fish (some are common aquarium fishes), finches, grasshoppers, and fruitflies, and the problem will be faced increasingly as many more groups of organisms are studied using evidence from multiple different genes.[13] To understand the problem better, we need to understand more clearly the difference between a gene tree and a species tree.

WHAT IS A SPECIES TREE?

Species trees are the types of evolutionary trees we are accustomed to seeing (Figure 2.3). In a species tree, the tips of the branches represent the different species that exist in the present day. In the case of the hominoid tree, there is a branch leading to chimpanzees, another branch leading to gorillas, and another leading to humans. As we have seen, up through the 1990s, the order in which these species branched from one another was still unclear. The actual branches of a species tree are "fat" because they represent populations evolving through time that contain many individuals. In the hominoid case, there is the population evolving along the chimpanzee branch, and populations evolving along each of the gorilla and human branches. The places on a species tree where branches diverge from each other are not discrete points in time. Instead, they indicate the cessation of gene exchange between two once-connected populations as they cease to interbreed. This process, known as speciation, can take many generations to complete.

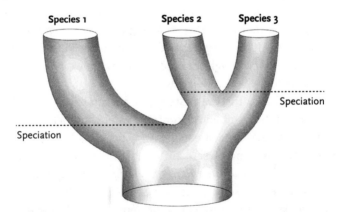

Figure 2.3: A species tree represents the true evolutionary relationships among three different species. Species 1 branched off (or speciated) at a time prior to when species 2 and species 3 speciated from one another.

A very important aspect of a species tree is that each of the many different individuals making up the branches carries many thousands of different genes. Each of these many genes has an evolutionary history represented by its own tree, called a gene tree. Thus, a species tree contains an amalgam of many, many thousands of evolutionary trees of different genes. Moreover, the way that any single gene tree branches, as long as it is far enough apart from another gene on the DNA strand, can in theory be very different from the other gene and can be different from the species tree.

WHAT EXACTLY ARE GENE TREES?

When the first human genome was sequenced in 2001, one of the great surprises, as mentioned earlier, was that our genome contained so few genes—the most recent estimates are around 21,000. Most primate genomes, including those from the hominoids, contain a similar number of genes. The majority of these genes share a one-to-one evolutionary correspondence, meaning they were inherited from the same common ancestor in the distant past. Genes from different species that share the same evolutionary origin are referred to as homologous genes. Within the genomes of humans, chimpanzees, gorillas, and orangutans, for example, there is a common gene called *beta globin* (HBB) that codes for one subunit of the hemoglobin protein, which transports oxygen around the body. (Note this is one specific gene of the set of six genes in the beta globin cluster, mentioned previously.) To make a valid comparison of this gene among species, with the aim of building an evolutionary tree, we must first isolate homologous *HBB* genes from the genomes of different species and then determine the sequence of hundreds of nucleotides that make up this gene. Using computer programs, we can now build a tree known as a *gene tree* that shows the evolutionary relationships among the different versions of the *HBB* gene from the different species. The tree describes the evolutionary relationships among the different versions of the gene present in the genomes of the different species. It is of crucial importance to emphasize that *the tree represents only the evolutionary history of this particular gene*. It is not the evolutionary tree of the species, a fact that has caused much confusion not only among nonspecialists but also among researchers.

Let's look at the tree for the hypothetical gene 1 in Figure 2.4. The very bottom of the gene tree represents the ancestral gene copy from which all present-day gene copies—those found in the three different species—have descended. The present-day copies of gene 1 are found at the tips of the branches. As we move upwards from the base of the tree and as branches

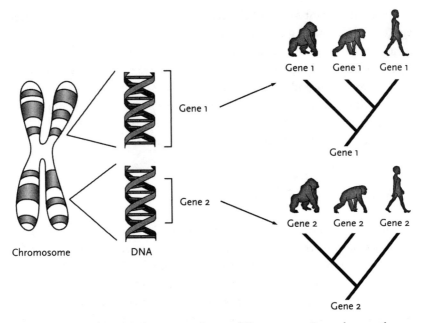

Figure 2.4: Representation of gene trees for two different genes. For each gene, the versions of the gene in three different species are being compared. The branching of the trees indicates the branching of the gene in evolutionary time. Notice the branching for gene 1 and gene 2 are different.

begin to emerge—first one branch and then two more branches—we are seeing a depiction of the evolution of this particular gene through evolutionary time. The lines of the tree represent gene lineages, or the persistence of a gene through a long chain of replication events as it was passed forward in time from parent to offspring over many generations. The places on the gene tree where one gene lineage has split to form two descendant lineages represent DNA replication events in which each of the descendant gene lineages was inherited by different species.

So far we have described a gene tree travelling upwards from its base to the tips of its branches. However, looking again at Figure 2.4, we can also trace gene copies backwards in time, starting with the tips of the branches and going back to the points at which each gene copy meets up with another gene copy, and finally going all the way to the base, where there is only a single ancestral gene copy. We call the process of merging between two gene lineages *coalescence*, like two streams traced backwards to the single river from which they both originated. For the homologous gene copies from three different species, it will always be the case that two of these copies will coalesce first, forming the single ancestral gene copy of these two gene copies. Then moving backwards, this ancestral gene copy will ultimately coalesce with the gene from the third species. When this happens, we say that the homologous gene copies from the three different species have coalesced back to their single common ancestral gene copy.

The 21,000 genes in our genome represent only about 1.5% of our genome. As you recall, for much of our genome there simply are no genes. Still, even for regions of our genome where there are no genes, it is possible to isolate and compare the homologous DNA from different species. In these cases, even though genes are not technically being compared, the shorthand way to refer to an evolutionary tree built from these DNA regions is still a "gene" tree.

HOW CAN GENE TREES BE DIFFERENT FROM EACH OTHER?

That gene trees might differ for a set of species, as do the trees for gene 1 and gene 2 in Figure 2.4, is due to the clustering of genes onto different chromosomes. There are twenty-three pairs of chromosomes in humans, and twenty-four pairs in each of the three great apes. In the process of sexual reproduction, offspring inherit one half of each pair of chromosomes from one parent and the other half from the other parent. Which half of the pair comes from which parent is a matter of chance. You can easily observe how this works in families where there are numerous children. As a simple example, one parent may have blond hair and blue eyes and the other might have dark brown hair and brown eyes. Although the inheritance of hair and eye color is complex and involves multiple genes, it is safe to say that if the parents have multiple children, some of the children will have a combination of features that neither parent has: brown hair with blue eyes, or blond hair with brown eyes. The reason for these new combinations of traits is that some genes coding for hair color and eye color are found on different chromosomes, and whether these different chromosomes are inherited together or not is a matter of chance. Because the random process of chromosome assortment is magnified over the many generations of evolutionary time, it means that genes from different chromosomes can have different evolutionary histories, or gene trees.

Even genes located on the same chromosome can have different evolutionary histories (as in Figure 2.4) if they are far enough apart from each other on the chromosome. In the process of sexual reproduction, a region of a chromosome can and does break off from its original chromosome and then is exchanged with the equivalent region on the other chromosome of their pair, a process known as crossing over (Figure 2.5). Through this process, the original combination of genes on a chromosome can be

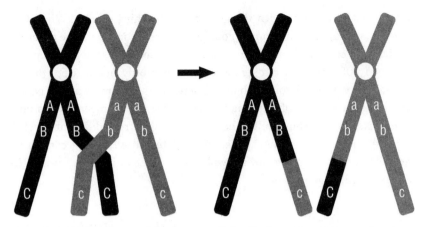

Figure 2.5: Crossover between chromosomes during the formation of sperm and eggs.

shuffled into novel combinations in different offspring. (There needs to be a distance of nearly 25,000 bases between two genes for the crossing over process to break them up.) If genes are closer together than this, it becomes increasingly unlikely that they will form new combinations in the offspring. Genes that are close together and always appear in the same combination are said to fall within the same linkage group. In evolutionary terms, genes located within the same linkage group will share the same evolutionary history and thus will have identical gene trees. Genes within different linkage groups can have different evolutionary histories (or gene trees) because these genes can recombine by crossing over.

When I was conducting postdoctoral research on human evolution with Jody Hey, a population geneticist at Temple University, I remember him referring to a species trees as a "fat tree" and their gene trees as "skinny trees." As you can see in Figure 2.6, the gene trees appear inside the species tree. In the illustration, we are assuming the branching order of species, but in reality the actual species tree remains unknown until many gene trees are determined. The gene tree for gene 1 is the skinny tree that appears inside the species tree, and it is represented by a solid thin line. If you examine the branching pattern of the gene tree, you will see that it exactly matches the branching pattern of the species tree. For this gene, we have species tree-gene tree congruence. When we examine the tree for gene 2 drawn in a thin dotted line, however, we can see that its branching pattern does not match the branching pattern of the species tree, a case known as gene tree-species tree incongruence. Now take a look on the right side of the figure, where we have added a third gene tree to the species tree. This gene tree (in gray) also does not match the branching pattern of the species tree and

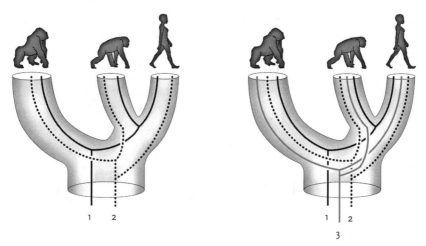

Figure 2.6: On the left, the species tree ("fat tree") for the African great apes and humans shows the close relationship between humans and chimpanzees. Inside the species tree are two thin gene trees, one that matches the species tree (gene 1, solid line) and one that does not (gene 2, dashed line). On the right, the same species tree is shown with a third gene tree (gene 3, gray line) that also does not match the species tree.

has a different branching pattern from either gene tree 1 or 2. Such gene tree-species tree mismatches can happen for any closely related species. It is also the situation for the African great apes and humans where different genes have trees that follow all three possible different branching arrangements, with some congruent with the species tree and others not.

If we were able to fit more gene trees in our illustration—for it is possible to build hundreds of thousands of gene trees—we could represent them all as skinny trees within the fat species tree. Since we know that only a single species tree represents the order in which the species separated from each other over evolutionary time, how is it that genes can have different evolutionary relationships than the species from which they were isolated?

One explanation for the variance between the gene trees within a species might be the pressures of natural selection, the mechanism of evolution proposed in 1859 by Darwin, in which individuals who have evolved and inherited a beneficial feature (or gene) have a higher probability of survival and reproduction than individuals without this feature. The process of natural selection might bias a gene tree, so that homologous genes in unrelated species evolve in similar ways. This makes the genes (and the species) appear to be closely evolutionarily related even though they are not.

We can see an example of the power of natural selection to shape the evolutionary trees of particular genes in the enzyme known as lysozyme that virtually all animals produce (in tears, saliva, and other body fluids), which plays a protective role by generally degrading bacteria. Because the lysozyme present in both ruminants like cows and sheep and the southeastern Asian monkeys known as langurs share similar properties, such as a capacity to function in acidic environments and a resistance to the stomach enzyme pepsin, in 1987 the geneticists Caro-Beth Stewart at University of Albany and the late Allan Wilson investigated the molecular basis for these similarities by determining the amino acid building blocks of the enzyme. In a remarkable finding, they discovered that lysozyme in these two unrelated mammals evolved many of the same unique amino acids, and that these changes were most likely the basis of the enzyme's similar properties. The evolution of two distantly related animals in the same way is called evolutionary convergence, and in this case, the lysozyme of cows and langurs had converged so as to function efficiently in foregut fermentation in both species. Furthermore, when Stewart and Wilson built an evolutionary tree based on the amino acid sequence of lysozyme in a variety of animals—including langur and cow, but also horse, mouse, baboon, and human—they found the tree to be very different from the known species tree.[14] In the lysozyme tree, langurs appeared more closely related to cows than to other primate species, like baboons and humans!

Despite the power with which natural selection could potentially reshape gene trees, however, there is little evidence that it has influenced the shapes of trees of the thousands of genes across the genome. The type of natural selection that could potentially reshape a gene tree is thought to be quite rare in the genome. This might seem surprising however since we are so used to thinking natural selection has the potential to change almost anything. Moreover, since most genes across the genome have very different functions, even if natural selection has altered the tree of one particular gene, it is not very likely to have altered the tree in exactly the same way as it has for another gene.

Because we believe that there is only a paucity of gene trees reshaped by natural selection, population geneticists have attributed most gene tree-species tree incongruence to another evolutionary process known as *random* lineage sorting. But how can gene trees be so different merely by chance? Keeping in mind the concept of coalescence, where we trace homologous genes from different species back to the point where they join, we can see that for each gene tree, there are two coalescent events occurring successively back in time. If we take gene tree 1 as an example (Figure 2.6, solid line), we see the first gene coalescence occurs just prior to the speciation between chimpanzees and humans, and within their common ancestral population. At this point there are two gene copies remaining—the homologous gene in the present-day gorilla, and the common ancestral gene in the ancestral chimpanzee-human population. These two homologous genes coalesce towards the bottom of the tree, within the ancestral population for all three species, forming a single ancestral copy of gene 1.

In contrast, in gene tree 2 (dotted line), we see that none of the three homologous genes coalesce in the chimpanzee-human ancestral population. In fact, we see that all three of the genes coalesce very closely in time within the common ancestral population for all three species. This suggests that within the ancestral population, there were already three different copies of gene 2 and individuals within this population differed from one another by having slightly different versions of gene 2 within their genomes. This is no different than what we see in human populations today, where different people have slightly different forms of genes influencing eye color or hair texture. When homologous genes coalesce in the ancestral population of three species, it becomes a matter of chance as to which of the three species' lineages will inherit which gene copy. For genes that coalesce so deeply in evolutionary time, three possible gene trees are equally probable.

We now can understand in evolutionary terms how the hominoid trichotomy problem arose. It is due to the fact that some gene copies have ancient coalescences and to the random way they consequently form trees. But could examination of gene trees help to solve the problem? Certainly several scientists felt it was worth considering alternative hypotheses of speciation. Instead of insisting on speciation as a two-way bifurcation process, where only two species emerge from a common ancestral species, perhaps the process was really a three-way split. Perhaps the ancestral species of the African great apes and humans was one large population spread out across equatorial Africa, and the three species—the gorilla, chimpanzee and human—evolved almost simultaneously from this widespread ancestral species. Genetic evidence from baboons, macaques, and other widespread primate species suggests that such complex speciation could theoretically occur, but was there really a three-way evolutionary split for the hominoids?

Random lineage sorting, however, is expected to cause gene trees to differ from the species tree for a much larger fraction of genes. Random lineage sorting is not dependent on a particular gene's function. Instead, it is dependent on random evolutionary processes that operate on every gene all the time. But there are two questions we need to ask about the evolutionary past to get an idea of just how many genes have experienced random lineage sorting. First, how big was the ancestral population just before it underwent the speciation process? Second, how much evolutionary time elapsed between speciation events in the species tree?

In the two species trees below (Figure 2.7), we can see how these two aspects of evolutionary history can influence how common random lineage

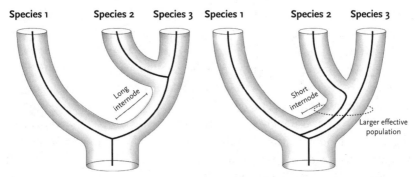

Figure 2.7: Two species trees with identical branching patterns. However, in the species tree on the left, the size of the common ancestral population from which species 2 and 3 evolved is relatively small and there is a long period of time (internode) between when species 1 branched off and when species 2 and 3 separated from each other. In the species tree on the right, the internode is comparatively shorter and the ancestral population size is larger, increasing the probability that the gene tree will not match the species tree.

sorting will be. The first species tree represents a small ancestral population (for species 1 and 2) and a long internode, or time interval between speciation events. In a species tree with these two properties, the likelihood is greater that the homologous gene copies from species 2 and species 3 will have coalesced within the interval of time between the two speciation events, resulting in a gene tree that perfectly matches the species tree. In general, the probability that a gene tree matches the species tree is dependent on how long the time interval (or internode) is between speciation events with respect to the size of the ancestral population. The reasons stem from the fact that coalescence is a random process, and the chance simply becomes greater, when the internode is longer, that the two gene copies happen to coalesce together during the internode and not at a more ancient time in the past. The random nature of the coalescent process also explains why population size is important. When a population is small there are fewer people carrying genes in the population, and it is more likely that their genes coalesced together in the recent past. And, when the gene copies from species 2 and species 3 coalesce within the internode, the gene tree will match the species tree.

The second species tree in Figure 2.7 shows a very different situation. Here the common ancestral population of species 2 and species 3 is larger compared to the time between speciation events, which is now very short. The short time between speciation events reduces the probability that the two genes from species 2 and 3 will have had time to coalesce during the internode. Instead, it increases the probability they coalesced earlier in the common ancestral population of all three species. The larger size of the population during the internode means that any two randomly selected individuals will be genetically more dissimilar to each other compared to the case in a smaller population. In large populations then, the gene copies carried by different individuals are more likely to coalesce prior to the internode, and within the common ancestral population of all three species. As we saw earlier, when the gene copies from the three species coalesce only within the common ancestor of the three species, it is a matter of chance as to which gene copy will be inherited by which of the three present-day species. This increases the probability that the gene tree does not match the species tree, as is the case in the second tree in Figure 2.7.

RESOLUTION

This greater understanding of the basis for gene tree incongruence would soon lead to a breakthrough in the hominoid trichotomy problem. In 1997, Maryellen Ruvolo applied population genetics theory about gene tree-species tree mismatch to new DNA sequence data sets for the great apes and humans.[15] In particular, she applied a statistical test devised by Chung-I Wu,[16] an evolutionary geneticist at University of Chicago, that takes into consideration the numbers of gene trees available for a group of species, and evaluates whether the most common gene tree occurs frequently enough to indicate the actual species tree.

To understand the basis of this test, remember that a species tree represents the evolutionary history of populations that have become reproductively isolated during the speciation process. This is a process whereby populations become gradually reproductively isolated from one another and no longer exchange genes. Reproductive isolation means that the thousands of genes in the genome of individuals on one branch of the species tree are isolated from all the thousands of genes in the genome of individuals on the branch from which it diverged. That is, speciation events (or branching regions) of a species tree will have also caused gene trees to have branching events that coincide with the branches of the species tree. Thus, the species evolutionary history should be reflected in the majority of the genes of the species. As we have learned, random lineage sorting can cause a minority of genes to have gene trees that do not reflect the species relationships. Wu's test acknowledges that some gene trees will differ from the species tree simply by chance. It then calculates the probability that each gene tree represents the actual species tree, based on how frequent this gene tree is within a distribution of all the gene trees known.

But before Ruvolo could apply the test, there was one requirement she needed for each gene tree. Each gene tree had to have come from genes from distinct linkage groups within the genome, meaning they had to have been inherited independently of each other. The first step in Ruvolo's analysis was therefore to determine if the different hominoid gene data sets came from different linkage groups. If two genes were closer than 25,000 base pairs on a chromosome, they were grouped together as a single genetic unit for analysis and represented by a single gene tree. The multiple genes from the mitochondrial genome also were grouped together into a single linkage group. The circular mitochondrial genome is only about 16,500 base pairs in length and contains a total of 37 genes. It does not undergo recombination, meaning there is no crossing-over process in its inheritance as there is for most of the nuclear genome.

After grouping all genes into separate linkage groups, Ruvolo was able to build fourteen well-supported gene trees. As seems appropriate for such a long-standing conundrum, the various gene trees suggested three different possibilities. Eleven different trees showed chimpanzees and humans to be the most closely related of the three, two trees grouped chimpanzees and gorillas together, and only a single gene tree paired humans and gorillas together. For the fourteen gene trees constructed, Wu's test requires that ten gene trees be identical to assume with high probability that they indeed coincide with the species tree. Since eleven gene trees showed chimpanzees and humans as most closely related, there was sufficient statistical support to accept this gene tree as the most probable representation of the actual evolutionary relationships of the species. The hominoid trichotomy problem had been resolved, with chimpanzees (and their close cousins, the bonobos) determined at last to be our closest ape relative.

Today, DNA technology and methodology have advanced to a stage where it is relatively straightforward as well as financially feasible to fully sequence entire genomes. Indeed hundreds of individual human genomes have now been fully sequenced, and we have full genome sequences for the chimpanzee the bonobo, the orangutan, and the gorilla.[17,18] We also already have full genomes from multiple individuals from each ape species, and the species tree built for all these many genomes indeed confirms humans and chimpanzees as most closely related.[19] With all these full genomes available, researchers have already been able to see a much larger set of trees within the forest—thousands of gene trees within the species tree.

THE GENOMIC REVOLUTION: THE CLOUD FORESTS

But as is usually the case when we are considering our evolutionary history, solving one puzzle only tends to lead us to the next. "Phylogeny is more like a statistical distribution than a simple tree of discrete thin branches," the evolutionary biologist Wayne Maddison at the University of British Columbia wrote in 1997. "It has a central tendency, but it also has a variance because of the diversity of gene trees. Gene trees that disagree with the central tendency are not wrong; rather, they are part of the diffuse pattern that is the genetic history" (p. 534).[20] Maddison has described phylogenic research as "a cloud of gene histories" with an inherent "fuzziness" to them. Likewise, in his book *Genes, Categories and Species* (2001) Jody Hey has also described species like clouds with diffuse boundaries.[21] Now that we have sequenced entire hominoid genomes, what can we say about their central tendencies, and what can we say about how much cloudiness surrounds ape and human evolutionary history?

Not surprisingly, having analyzed the full genomes of humans, chimpanzees, and gorillas and built gene trees across thousands of separate linkage groups, we've found that the human species appears to be most closely related to chimpanzees across roughly two-thirds of our genome.[22,23] But for a remaining third of our genome, the data reveal that we are not the chimpanzee's closest cousins. For these genes, humans are either genetically more closely related to gorillas than to chimpanzees or chimpanzees are genetically more closely related to gorillas than to humans. You can see this "cloudy" evolutionary history of our genome in the illustration below of human chromosome number 1 (Figure 2.8), lying on its side. For many of the locations along this chromosome, our DNA is most closely related to chimpanzees (shown below the chromosome), but for many other regions along this chromosome our DNA is more closely related to the gorilla, and in still other locations our DNA is not especially close to either chimpanzees or gorillas (shown above the chromosome). In fact, a very similar pattern is true of all the other chromosomes in our cells.

Now that the bonobo and orangutan genomes have been fully sequenced, we can also build gene trees that compare humans with each of these ape species and allow us, in turn, to estimate the proportion of our genome that is more closely related to these apes. Though bonobos are the nearest ape cousins to chimpanzees, having separated from them only about two million years ago, it is striking that 3% of our genome is more closely related to the bonobo's genome than to the genome of chimpanzees.[18] Orangutans are an even more distant ape relative of humans, having diverged from our lineage more than thirteen million years ago.[24] Despite this long time, it is estimated that for 1% of our genome we are more closely related to orangutans than to chimpanzees, bonobos, or gorillas.[25] Although 1% may not seem like a lot of DNA, in fact it amounts to thirty million nucleotides of orangutan DNA. Indeed, it is possible that those regions of our genome in which we are more similar to one or another of the great apes might be

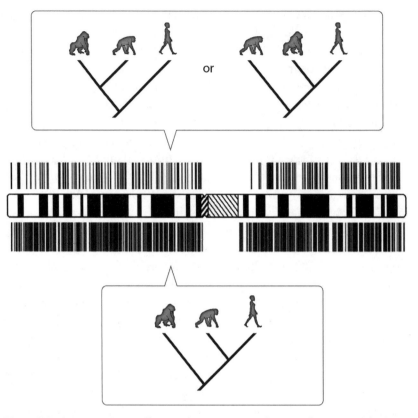

Figure 2.8: A representation of human chromosome 1 is shown in the center of the figure. On the lower side of the chromosome are indicated all those specific locations where human and chimpanzee genes have DNA that is most closely related. On the upper side of the chromosome, two other DNA locations are indicated, where human DNA is more closely related to gorilla DNA than to chimpanzee DNA, or where chimpanzee DNA is more closely related to gorilla DNA than to human DNA. All human chromosomes have a similar mosaic pattern of ancestry. (After Ebersberger, et al., 2007.[11])

important functional regions of the genome that provide the DNA blueprints for certain anatomical or behavioral features. We also may be unspecialized, unlike any of these other apes but more like the primitive ape from which all living great apes evolved. Our genome contains the seeds of many unique features, including many of our well-known adaptations like erect posture and bipedal locomotion, our complex language abilities, and our massive brains and greatly augmented cognitive abilities.

The core of Darwin's argument over 150 years ago in the *Origin of Species* was that all living organisms on Earth are connected by a series of common ancestral species as we go back in time. Although he did not know about genes, he did believe in a genealogical connection among all life forms and particularly remarked on our close evolutionary relationship to the African great apes. Today, in a way that Darwin could never have imagined, we can dramatically see our evolutionary connection with the other apes reflected in the evolutionary trees we build for the many thousands of "genes" that make up our genome and the genomes of the apes. It is clear that we share the largest portions of our genome with chimpanzees (and bonobos), more so than with any other ape. However, for thousands of genes, we are now able to see the overwhelming evidence of our genomic continuity with all the great apes. This evolutionary continuity has emerged out of a process whereby genes in the common ancestors we share with gorillas and orang-utans sorted randomly into the genomes of the living great apes. Within our genome, we can see not only our inner chimpanzee, but our inner bonobo, gorilla, and orangutan!

CHAPTER 3

The Great Divorce

How and When Did Humans and

Chimpanzees Part Ways?

Mismatches between gene trees and the species trees are rampant among the great apes and humans, and for a good while this confused our understanding of our closest evolutionary relationships. We might be tempted to construe such conflicts between gene trees and the species tree as evolutionary "trickery" and a royal nuisance, what we commonly call the gene tree-species tree problem. This is an appropriate assessment for scientists whose paramount concern is to discover the order in which species separated from one another over evolutionary time. However, when we look through the lens of population genetics, the extent to which gene trees conflict with the species tree gives us important clues about the evolutionary past. It might seem counterintuitive, but the confusion actually helps us to see our past more clearly.

As we've seen, we can trace copies of homologous genes isolated from the genomes of different species in the present to the points where they join together (or coalesce) in the past. One of the main factors causing any single gene tree to be different from the species tree is that gene copies drawn from the genomes of two different species coalesce so deeply in the past that this coalescence doesn't occur in the last common ancestor of these two species. In chimpanzees and humans, for example, this may mean that their genes coalesce in the common ancestor that chimpanzees and humans shared with gorillas.

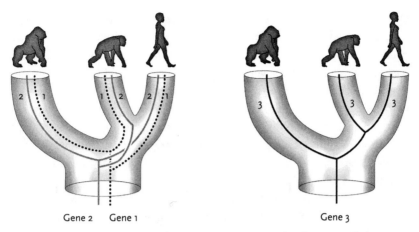

Figure 3.1. The species tree on the left shows two gene trees that do not match the species tree. For these gene trees, the order by which their separate branches emerged does not match the order by which the species separated from each other. On the right, however, the gene tree shows the same branching order as the species tree.

When coalescence occurs deep in the past, the gene copies can be related in random ways. The gene trees in Figure 3.1, trace the coalescence of gene 1 (dotted tree), gene 2 (gray tree), and gene 3 (solid black tree) backwards in time. We see that that the copy of gene 1 from the chimpanzee coalesced with the gene copy from the gorilla before eventually coalescing with the human copy of gene 1, producing a gene tree (dotted tree) different from the species tree. Likewise, gene 2 also coalesced very deeply within the common ancestor of the African great apes and humans. In this case, the copy of gene 2 from the human genome coalesced with the gene copy from the gorilla genome before coalescing with the gene copy from the chimpanzee genome. Genes 1 and 2, therefore, have gene trees that belie the real species relationships—the bane of any phylogeneticist. On the other hand, the copies of gene 3 from chimpanzees and humans do not coalesce as deeply in time, but merge instead within the common ancestor shared by these two species. Only further back in time do they coalesce with the copy of gene 3 from the gorilla. Therefore, when genes have a more shallow coalescence they tend to match the species tree. Remember, there are two factors that increase the likelihood of this happening; one, if there was a relatively long time between when gorillas first branched off and the time when humans and chimpanzees branched from each other; and two, if the population size of the ancestral species that split to eventually evolve into humans and chimpanzees was relatively small.

When we look at the thousands of genes and different linkage groups (separate recombining regions) within the genome we can see large differences in the times at which they coalesce in the past. For some species, almost all genes may coalesce in a very narrow window of time in the past. For other species, there may be a very large window of time in the past, ranging back over several millions of years, in which different genes may coalesce. For chimpanzees and humans, there is a very large window of time over which gene copies from the two species coalesce together, with the deeper coalescing genes being prone to gene tree-species tree mismatch. For evolutionary geneticists, however, the confusing proliferation of gene trees for great apes and humans can actually clarify our shared evolutionary past with the great apes.

For one thing, we can get a better handle on the time—millions of years in the past—when the common chimpanzee-human ancestral population split to form two separate evolutionary lineages leading to humans and chimpanzees (and bonobos). We can also estimate how big or small their common ancestral population was. In fact, we can go further back in time to estimate the sizes of other populations, like the common ancestral population that humans and chimpanzees shared with the gorilla. Gene tree-species tree mismatch also allows us to examine questions about the nature of the speciation process that produced the human and chimpanzee lineages. For example, how fast or slow was their evolutionary separation? And what was the cause of their evolutionary separation? As one possibility, many anthropologists believe that the common ancestral population of chimpanzees and humans occupied habitats that ranged from tropical forests to woodlands, much like the habitats in which common chimpanzees and bonobos live today. Did a series of genetic mutations occur between two groups of individuals within this ancestral population that gradually prevented them from interbreeding, leading to their separation? Or did a geographic divide drive groups of individuals in this common ancestral population apart, eventually isolating them into two separate species?

HOW BIG WERE ANCESTRAL POPULATIONS?

While it might seem counterintuitive from the perspective of the vast size of human populations today, there is some intriguing genetic information indicating that ancestral population sizes of the human lineage were quite small, especially compared to the size of the human-chimpanzee ancestral population. And as we'll see, knowing the size of the human-chimpanzee ancestral population gives us a clue as to why we suffered the hominoid trichotomy problem for the apes for so long.

To measure the size of a population or species, it is theoretically possible to count every living member and come up with its census size. However, our census size will likely be greater than the number of individuals that actually bred and contributed genes from generation to generation. For example, in some species a minority of males may breed with all the females in the group. In such a species, these males contribute disproportionately to the pool of genes passed down to the next generation and the males that didn't breed are essentially irrelevant from an evolutionary genetics standpoint. The number of breeding individuals is what geneticists wish to know because this number can give a more realistic interpretation of how evolutionary forces acted on the species as it evolved through time. Thus, from a genetics perspective, the population is effectively smaller than the census size and this measure of population size is referred to as the *effective* population size. Effective population size is really an indication of the number of individuals that participated reproductively in the evolutionary past and were lucky enough to pass their genes forward to descendants that are alive today.

So what is the effective population size for the human species today? Well, it likely will sound surprising, given the vast size of the human population, but genetic studies have fairly consistently estimated it to be around 10,000.[1,2] It is crucial to recognize the vastness of human evolutionary time since our separation from chimpanzees—five to six million years—and that our massive population explosion has only occurred very recently in human evolution. Genetic evidence suggests that human population size started increasing only around 50,000 years ago.[3] But the most dramatic increases have only occurred since the emergence of agricultural and animal domestication around 11,000 years ago[4,5] and more recently since the start of the Industrial Revolution in the eighteenth century. Thus, for most of our evolutionary history since we separated from chimpanzees, it appears that the size of the population of human ancestors who bred and left genetic descendants was modest to small.

What can we tell about actual numbers of people in ancient populations from the effective population size? Although the relationship between effective population size and the actual number of individuals in the population is quite complicated, depends on many immeasurable factors as well as fluctuations in population size over evolutionary time, one admittedly rough rule of thumb that has been commonly used says that effective population size is about one-third of the census size of the population. This would put our ancestral population size at about 30,000 individuals. There is much debate on this issue, however, and some geneticists have estimated the census size to be considerably larger. Nevertheless, one of the striking things that recent evolutionary genetics has discovered is that the common ancestor of humans and chimpanzees had an estimated effective population size considerably larger than humans do, with estimates ranging up to five to ten times larger.[6] This is significant, because as we will discuss in chapter 4, population size reduction appears to have had important consequences for the trajectory of our evolution as well as for human health and disease. How do we determine the effective population size of our common ancestor with chimpanzees? And how can gene tree-species tree mismatches inform us about this population size?

COUNTING GENE TREES

As we've seen, two factors make it more or less likely that a gene tree matches the species tree. The first factor is the length of the time between two successive speciation events (Figure 3.2); for example, the time between the divergence of the gorilla lineage and the more recent time marking the divergence of humans and chimpanzees. The time between the two speciation events (or the internode) is determined by subtracting the time of speciation between humans and chimpanzees from the earlier time of speciation when the gorilla branched off. The second factor is the effective size of the ancestral population from which humans and chimpanzees later evolved. It is the relationship between these two variables that is important here, and can result in a low probability that a given gene tree

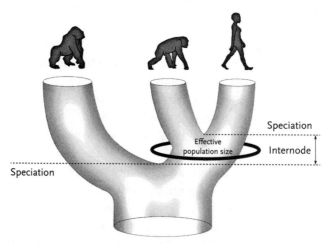

Figure 3.2: The probability that a gene tree will not indicate that chimpanzees are human's closest relative is dependent on two factors: the effective population size of the common ancestor of humans and chimpanzees (encircled) and the length of time, known as the internode, between the time when gorillas branched off and the later time when chimpanzees and humans separated.

will match the species tree. When the internode is small and the ancestral effective population size is large, then the probability of a gene tree not matching the species tree becomes greater.

In 1988, the population geneticists Pekka Pamilo and Masatoshi Nei developed a mathematical formula showing how the proportion of gene trees matching the species tree is influenced by the two factors above (length of internode and effective population size).[7] If this proportion of trees is known, then it is possible to obtain an estimate of the effective population size before we split from chimpanzees. The formula was applied to the hominoid trichotomy problem as soon as multiple DNA sequences from different genes had been determined for the great apes and humans.[8,9] To estimate the effective population size in the common ancestral human-chimpanzee population, we need two pieces of information. First, we need to count the number of gene trees that match the species tree, as well as the number of gene trees that do not match the species tree. Second, we need estimates of the time (millions of years ago) when humans and chimpanzees split from each other, and of the earlier time when gorillas branched off.

In 2001, molecular biologists Feng-Chi Chen from the National Health Research Institutes and Wen-Hsiung Li from the University of Chicago, analyzed eighty-eight DNA regions from humans, chimpanzees, gorillas, and orangutans.[10] Since a fairly large subset of the DNA regions in the Chen and Li study—thirty-six of the eighty-eight—were so similar (or even identical to each other) among the three species, they showed a three-way split among humans, chimpanzees, and gorillas. These DNA regions cannot be used in the analysis since their gene trees do not pair together any two species as separate from the third species. This fact is extremely interesting because it indicates just how genetically similar we are to our African ape cousins. Only genes that showed a bifurcating (two-way) branching tree—a human-chimpanzee pairing, a chimpanzee-gorilla pairing, or a human-gorilla pairing—were used in the analysis. Of the fifty-two trees, thirty-six showed a human-chimpanzee pairing whereas the remaining trees showed the alternative pairings. Using this information, the likelihood that a gene tree would match the species tree was estimated to be 36/52 or 69%. This percentage was the first crucial piece of information needed to plug into the population geneticists' formula for determining effective population size. (Whole genome studies since have estimated this percentage to be 70%, a nearly identical finding.)

The second piece of information needed in the analysis is the length of the internode. To find this, we need estimates of the dates when human and chimpanzees evolutionarily separated from one another, and the prior date when gorillas branched from humans and chimpanzees. Luckily, these dates can be obtained from the DNA regions of the three species by first counting the numbers of DNA differences between them and then applying a molecular clock-based method. Based on empirical data from many DNA data sets, the molecular clock holds that differences between DNA regions of different species accumulate steadily since the time they last shared a common ancestor. To estimate when the African great apes and humans separated from one another, the researchers counted the number of DNA differences between each species pair for each of the fifty-two genes. They then averaged these numbers across the fifty-two different genes, coming up with a single number representing the amount of DNA differences between the species, and found that the average number of DNA differences between humans and chimpanzees was 1.24 per 100 bases (Figure 3.3). Between humans and gorillas, it was 1.64 per 100 bases. (In the most recent whole genome studies the numbers are 1.37 and 1.75, again very similar!) Rounding these numbers a bit, there is only about one DNA difference in every 100 DNA bases between us and chimpanzees, and less than two DNA differences per 100 DNA bases between us and gorillas (stimulating cocktail party facts!). On our tree, these values have been halved because we assume that half of the DNA differences between two species arose during the evolution of one species and the other half of the DNA differences arose during the evolution of the other species.

We can convert these DNA differences into actual dates if we use a calibration reference point, a date from a relevant fossil ancestor determined using

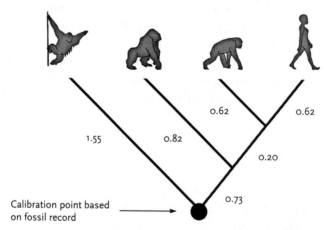

Figure 3.3: Evolutionary tree showing the number of DNA differences between the great apes and humans. For example, when differences are averaged across hundreds of bases compared between chimpanzee and human genes, there are approximately 1.24 DNA differences found for every 100 bases (Chen and Li, 2001). Approximately half of these differences (0.62) evolved along the human lineage and the other half evolved along the chimpanzee lineage.

geological methods. A commonly used calibration point for the great apes and humans is the estimated date for when orangutans branched off the ape evolutionary tree. We have a fairly good estimate of this time as a result of excavations by David Pilbeam at Harvard University and John Kappelman at University of Texas and their colleagues, who discovered fossil material of an ape from Pakistan known as *Sivapithecus*, an extinct ape that has features of the skull that closely resemble those of modern orangutans. Radiometric dates obtained for *Sivapithecus* indicate that it lived approximately 13 million years ago.[11] Since *Sivapithecus* was on a very early part of the branch leading to modern orangutans, it is likely that orangutans actually branched off somewhat before this. Therefore, dates of 14 million years or even slightly older are used for the calibration. Of course it is also necessary to count up the number of DNA differences between orangutans and humans. When Chen and Li counted up these differences across the gene regions, they found that there were 3.1 DNA differences on average between orangutans and humans (Figure 3.3).

We can now estimate mathematically the age when chimpanzees and humans split from each other and the earlier time when gorillas split from the human-chimpanzee lineage. (Don't hang your hat on these specific estimates, since shortly we will look at estimates from more recent genomic studies.) Using the DNA differences between humans and orangutans, we can establish how many DNA differences arise every million years. Doing the math, we find that 0.238 DNA differences per 100 bases accumulated between species every 1 million years (we call this the phylogenetic mutation rate). To determine when humans and chimpanzees split, we take the phylogenetic mutation rate and divide it by the total number of DNA differences between these two species (1.24) which gives us 5.2 million years for the human-chimp split. We can do the same thing to estimate when the gorilla branched off and get 6.7 million years. This gives us the crucial second piece of information we need, the length of time—or internode—between the two speciation events. This turns out to be 1.5 million years, subtracting 6.7 from 5.2 million years. While speciation dates for humans, chimpanzees, and gorillas have changed somewhat in more recent whole genome analyses, most studies conclude that there was a relatively short amount of time between the two successive speciation events. In fact, the internode time is only about 30% of the time that the human lineage has been separate from the chimpanzee lineage.

The internode time and the percentage of gene trees that match the species tree could now be plugged into the population geneticist's formula in order to estimate the size of the common ancestral chimpanzee-human population. Surprisingly, the common ancestral species was found to have a very large effective population size, ranging anywhere from 52,000 to

96,000, a surprising five to ten times the effective population size of modern humans.[10]

ENLARGING THE SNAPSHOT

In these early studies, researchers chose a limited number of DNA regions to analyze out of the many thousands of possible regions that could be studied. Inferior-grade laboratory technology was the limiting factor then, but accelerating advancements in technology now make it much easier to collect far, far greater amounts of DNA data. Most studies today analyze all the DNA sites in the genome and hone in on the many DNA differences that occur between the genomes of different species. Remember that the genomes of humans and chimpanzees differ by an average of only one DNA difference per one hundred bases.[12] All told, there are roughly 30 million DNA substitutions (different bases) between them, and twice as many substitutions between humans and gorillas. Researchers can now sort these DNA differences into one of three categories: DNA sites supporting gene trees that pair chimpanzees with humans; DNA sites pairing chimpanzees with gorillas; and DNA sites pairing humans and gorillas. As seen in Figure 3.4, each group of sites supports one of the three different gene trees possible for the three species.

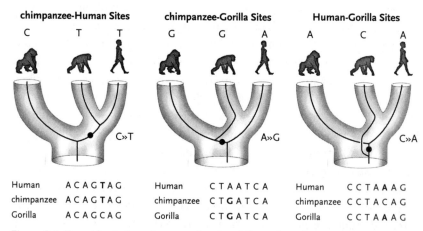

Figure 3.4: Three identical species trees, showing different branching patterns of the DNA tree within the species tree. The DNA change that supports each gene tree is shown to the right of each tree. The tree within the first species tree indicates that for this DNA site, chimpanzees and humans are most closely related since they both inherited a T DNA base, changed from the C DNA base in their common ancestor with gorillas. The tree within the second species tree shows that, for this site, gorillas and chimpanzees are most closely related because they inherited a G DNA base changed from an A DNA base. The third tree indicates that, for this site, humans and gorillas are most closely related because they inherited an A DNA base changed from a C DNA base.

What do these new genome-wide studies tell us about the effective population sizes of our shared ancestors? Since new studies compare billions of nucleotides among different species, much of the work done is very computer intensive and is managed by researchers in the new field of bioinformatics. Crunching the genomic data leads to estimates of effective population sizes very similar to the values calculated in early studies of limited numbers of DNA regions, revealing that even with a relatively limited amount of DNA sequences, evolutionary estimates can be quite accurate as long as these DNA regions come from independent (recombining) regions of the genome. These genomic studies estimate the effective size of the human-chimpanzee common ancestral species in the approximate range of 50,000 to 100,000.[6,13-15] If we use the admittedly imperfect one-third rule, described previously, this translates roughly into 150,000 to 300,000 individuals.

Recent genomic studies have been able to peer even more deeply into our evolutionary past. Using the orangutan and gorilla genomes completely sequenced in 2011 and 2012, respectively, studies have estimated the effective population size of the common ancestral species that humans and chimpanzees shared with gorillas, as well as that of the more ancient ancestral species shared with orangutans. These populations also appear to have been relatively large: 45,000 to 65,000 for the effective size of the human-chimpanzee-gorilla ancestral population as well as for the effective size of the ancestral population shared with orangutans.[6,14-16] Even though these numbers may change in new and increasingly fine-tuned analyses, overall genomic evidence indicates that our hominoid ancestors had considerably larger effective population sizes compared to that of the human species. As we'll see in the next chapter, humans appear to have undergone a fivefold to tenfold reduction in population size during their evolution.

WHEN DID THE GREAT EVOLUTIONARY DIVORCE HAPPEN?

How do we estimate when the human-chimpanzee split took place, using entire genomes? To begin, let's look at the two gene trees represented within the species tree for humans and chimpanzees (Figure 3.5). The speciation time represents the time when the two populations no longer produce offspring with each other and cease exchanging genes. For both genes 1 and 2, which have homologous copies in both chimpanzees and humans, we can trace their gene lineages back in time and observe that the two copies coalesce into a single gene copy in the common ancestral species. This is referred to as a coalescent event, and there is an ancestral copy of gene 1

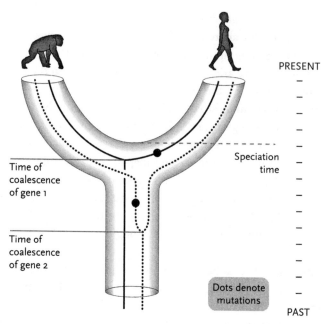

Speciation
time

Time of
coalescence
of gene 1

Time of
coalescence
of gene 2

Dots denote
mutations

PAST

Figure 3.5: This species tree shows the different coalescence times of two different genes. Gene 2 coalesces farther back in the past compared to gene 1. The dots on the gene trees represent DNA mutations.

and an ancestral copy of gene 2 in this species. If we now proceed forward in time, starting from the bottom of the tree, we see that the original single gene lineage for each gene splits into two copies. At the moment that this split occurs, there are two different copies of gene 1 and two different copies of gene 2. In fact, each copy soon becomes a distinct version of the gene, because of a mutation (the black dot) that occurred in one of the gene copies. The two different versions of the gene (also known as different alleles) are comparable to the different versions of genes we find today between two people which, for example, may cause them to have differently colored eyes. In this tree, we see that the present-day gene copies in chimpanzees and humans are descended from gene copy variants already present in the common ancestral population millions of years ago. In addition, some gene copies present in chimpanzees and humans today, which we compare when we build gene trees, had splitting times that occurred well before the two species themselves separated.

We can also observe in these trees, proceeding from the tips of the branches to their bases, that the coalescence time of gene 2 is deeper in the past than then the coalescence time of gene 1. This means that within the ancestral population, gene 2 split to form two distinct gene alleles well before gene 1. As described previously, when gene copies from humans and from chimpanzees coalesce so far back in time that they only coalesce within the common ancestral species shared with gorillas (not shown here), this can lead to gene tree-species tree mismatch. Until quite recently, it was common for researchers to estimate speciation times by simply determining the date a single gene split to form two different gene variants. This method overestimated the time of speciation, since gene copies always coalesce prior to the speciation event and, as we have just seen, sometimes much earlier. Today, we avoid this problem by analyzing the extent of variation in coalescence times among the many thousands of DNA regions of the entire genome.

The extent of variation in coalescence times among genes is directly related to the size of the ancestral population. When the various genes of two species are examined and found to coalesce at very disparate times in the past, this indicates the common ancestral population of these species was large. When many genes of two species are examined and all are found to coalesce at approximately similar times in the past, on the other hand, this indicates the ancestral population of the two species was much smaller. Indeed, one of the reasons we believe that the ancestral chimpanzee-human population was relatively large is that the thousands of genes compared between them coalesce at vastly different times. In fact, different gene regions are estimated to coalesce over a range of more than four million years, with many genes coalescing deeply in the common ancestral population shared with gorillas (Figure 3.6).[17]

So the question becomes how to arrive at a good estimate of the speciation time between humans and chimpanzees amid so much variation in gene coalescence dates. One way is to identify all DNA regions in the human and chimpanzee genomes that coalesced most recently in time, because the age of these regions would be closest to the time that humans and chimpanzees first started separating from each other. But how does one identify these genes. One method is to find only those human-chimpanzee DNA sites with the most recent coalescence dates since these are most likely to be near to the speciation time between humans and chimpanzees. These dates are then compared to the average coalescence dates for all DNA regions of the genome, no matter whether they are defined by human-chimpanzee, chimpanzee-gorilla, or human-gorilla sites. In a genome-wide study of thousands of regions across the genome by geneticist Nick Patterson at The Broad Institute of Harvard and MIT,[17] when just human-chimpanzee DNA sites were examined they were found to be approximately 15% more recent than the average coalescence of all regions across the genome (including human-gorilla and chimpanzee-gorilla sites). Using a calibration date based on the orangutan-human divergence, the study concluded that the most recent coalescence of gene regions between

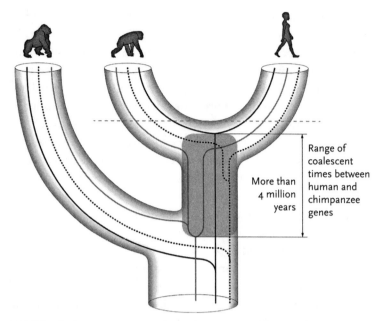

Figure 3.6: Inside this species tree are shown three gene trees (solid, gray, and dotted trees). In the solid tree, the human and chimpanzee gene copies coalesce very recently in time. In the gray tree, human and chimpanzee gene copies coalesce at a very ancient time in the past (note that this gene tree does not match the species tree). In the dotted tree, they coalesce at a time somewhere in between. Analyses of full genomes show that human and chimpanzee genes coalesce within a very large time frame of more than four million years.

human and chimpanzee genomes was around 5.4 million years ago. This indicates that humans and chimpanzees parted evolutionary ways somewhat more recently than 5.4 million years ago.

For the most part, the times of speciation, or separation, between humans and chimpanzees estimated in different studies of full genomes are fairly consistent with the date estimated above for much smaller data sets. Surprisingly, more than one study has placed the chimpanzee-human split even more recently in time, somewhere in the range of between about four and five million years ago.[14,16,18] Recent dates for their evolutionary separation are surprising because we have traditionally thought that more time would have been necessary for humans to accumulate their vast differences from chimpanzees and because, in some researchers' estimation, these dates disagree with fossil evidence. It is uncertain at present whether such recent dates based on the genome are accurate, but looking at them has raised the eyebrows (and the hackles) of some scientists who unearth fossils of ancient human ancestors. These fossil discoveries, as we will see, tend to push back the time of our origin more deeply into the past and as such present a conflict with the genome estimates.

Anthropologists have long speculated on how and why humans became reproductively isolated from their most recent primate relative. For example, was it a quick and clean split with no reproductive contact between the newly evolving chimpanzee and human species? Or was it a protracted and complex process with reproductive intermingling between the two emerging species—a slow and messy split? Genomic studies allow us to address these questions in new ways.

Ernst Mayr, a prominent evolutionary biologist of the 20th century, proposed that speciation most often occurs via the process he termed *allopatric* ("residing separately") speciation.[19] Allopatric speciation takes place when two subpopulations of a single species become isolated from each other, usually by some newly arising geographic barrier (e.g., a widening river, or the disappearance of part of a once continuous forest). Once the geographic barrier forms, interbreeding between the two subpopulations ceases and speciation occurs quickly. In human evolution, it has long been speculated that the earliest human ancestors moved into a dry grassland-savanna environment and became geographically, and soon reproductively, isolated from their forest-dwelling ancestors.

On the other hand, nature may never be so tidy. While the biological species concept states that "species are groups of actually or potentially interbreeding natural populations which are reproductively isolated from other such groups,"[20] this does not always bear out in reality. Indeed, there are quite a number of species living today that share a contiguous geographic border with another closely related species. At these borders where the two species meet, they interbreed and produce fertile offspring.

Anthropologist Clifford Jolly has studied such messiness among species in African baboons for over thirty years.[21] The five or so differentiated populations of common baboons each look quite distinct and occupy immediately adjacent and slightly different ecological zones within sub-Saharan Africa. For the most part, they are geographically isolated from each other and do not interbreed, leading some to call each a separate species. However, Jolly has found that almost all of them interbreed at hybrid zones where they meet, resulting in fertile offspring. These hybrid offspring then mate with either one of the parental species producing fertile offspring. Macaque monkeys across Asia are also known to freely interbreed where different species meet,[22] and there are many other examples in primates and in other animal groups. Of course, this raises the question of exactly how to define a species, which remains a conundrum for biologists.[23] Nevertheless, examples like this indicate that speciation is sometimes (perhaps oftentimes) a fuzzy affair where genes can be exchanged between incipient species for a considerable period of time before they become completely reproductively isolated from each other. Such gene exchange, or gene flow, is proposed to occur in the *parapatric* ("residing next to each other") process of speciation, where emerging species occupy adjacent areas but meet at zones where they hybridize. (Biologists also refer to another style of speciation called *sympatric* [residing together] speciation, where a single species splits into two different species even though they live in the very same habitat in the same geographic location.)

Do these different styles of speciation leave different clues in the genome? When we look at the DNA differences between the human and chimpanzee genomes, for instance, can we see a pattern that allows us to infer whether one or the other type of speciation was more likely? Researchers have not matched exact DNA signatures with these different forms of speciation, but there have been some theoretical analyses of the problem. Most of these analyses focus on distinguishing allopatric speciation from parapatric speciation since these are believed to be the most common ways species evolve.

Figure 3.7 shows two different hypothetical scenarios of speciation for chimpanzees and humans. The illustration on the left represents a hypothetical scenario in which speciation was quick, as in allopatric speciation. The illustration on the right represents a scenario in which the speciation process was prolonged, as in parapatric speciation. In the parapatric scenario, the gray brick wall represents the period of time when some genes were still exchanged through limited mating between the two separating species. The region where the brick wall becomes solid represents the time where interbreeding ceased and there was no longer any exchange of genes between the two species. Notice that the point of reproductive isolation, indicated by the upper dashed horizontal line, is exactly the same in both scenarios.

As we can observe, the only expected difference between the two species trees under the two scenarios is in the time for all gene copies from the chimpanzee to coalesce with gene copies from humans. In allopatric speciation, the time-frame within which genes coalesce is expected to be smaller than the time-frame within which genes coalesce under a parapatric mode of speciation. In other words, when genes are still being exchanged between species, even as the two species have already started separating from each other, the time-frame within which genes coalesce is greater.

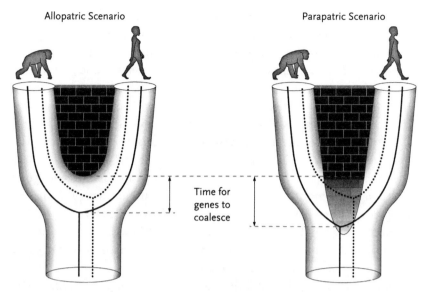

Figure 3.7: The species tree on the left represents a hypothetical scenario in which the evolutionary separation between humans and chimpanzees was relatively fast (as in allopatric speciation). The species tree on the right represents a scenario in which speciation was a prolonged process (represented by the slow "building" of a brick wall) because the two species continued to mate to limited degrees even as they were separating (as occurs in parapatric speciation). The process of building the brick wall indicates less and less gene flow until mating finally ceases at the completed brick wall. For each scenario, inside the species trees are found two different gene trees (dotted and solid). In the parapatric speciation scenario (on the right), as we trace genes back in time from their present copies, they are expected to coalesce over a greater range of times than in the allopatric scenario, where the range of coalescence times is considerably less.

A PROLONGED HUMAN-CHIMP AFFAIR?

The roughly four-million-year range of coalescence dates between human and chimpanzee genes leads us to believe the common ancestral population between humans and chimpanzees was large. In fact, given the large estimated population size, the University of Edinburgh population geneticist Nick Barton has determined that it would not be unusual to see genes from chimpanzees and humans coalescing over a time frame of over eight million years, even if the two species had a clean and quick evolutionary split.[24] According to this view, it would appear that an allopatric and relatively quick speciation model is consistent with the genomic evidence.

But were things really so simple? A lively debate on the matter is ongoing. The debate is no doubt motivated by the provocative explanation for the large range of coalescence dates given by the group of researchers who first detected it.[17] Patterson suggested that the wide range of coalescence dates could be explained by a very prolonged road to separation—that chimpanzees and humans initially separated slightly more than 6 million years ago and became reproductively isolated, but then about a million years later they began interbreeding again, before finally making the final split 5.4 million years ago.[17] This was surely a complex and provocative proposal!

The scenario of prolonged speciation with reproductive intermingling, published in 2006, received a great deal of attention in the press. One creative review article even dubbed the hybrids "chumanzees."[25] Of course, the idea of reproductive intermingling among humans and chimpanzees very early in their evolution would not be uncommon for two species emerging, even though the specific scenario of an initial separation, a very significant time gap, and then hybridization before a final farewell is far-fetched. Not unexpectedly, a number of critical publications followed pointing out the reasons why the genomes of the humans and chimpanzees do not support such a "long-and-winding road" scenario.[26-28]

However, let's keep in mind that humans and chimpanzees likely still looked very similar to each other in the very earliest stages after their evolutionary separation. In fact during this time they may have easily recognized each other as potential mates. In the age of genomes, new methods have been specifically designed to detect whether newly emerging species continued to interbreed even while in the midst of separating. In 2012, when these more subtle methods were applied to the genomes from the various great apes,[29,30] the results indicate that the two living gorilla species—eastern and western lowland gorillas—kept mating on and off for as long as several hundred thousand years before they finally split. The two living orangutan species were also found to have kept mating intermittently while they were on their way to becoming reproductively isolated. On the other hand, the split between chimpanzees and bonobos appears to have been a much cleaner and quicker evolutionary separation.[31] When the method has been applied to human and chimpanzee genomes, it has revealed that, as in orangutans and gorillas, our two species continued to interbreed intermittently for a period of several hundred thousand of years before becoming fully reproductively isolated from each other. So while it was not the long-and-winding road to separation between us and chimpanzees, it does appear that we dithered on, in a reproductive sense, before finally splitting.

As we've seen, according to the "savanna hypothesis," anthropologists have long speculated that the separation between us and chimpanzees was a clean break beginning with the movement of our earliest ancestors into the arid savannas in response to warming temperatures. The savanna is where we first became upright and bipedal and where, as the story goes, we eventually invented tools, weapons, and big brains. We have surmised that, since chimpanzees remained in the forests, our reproductive isolation came on quickly because our two preferred habitats were mutually exclusive. However, the newest genomic evidence suggests that as humans and chimpanzees were in the initial throes of separation, we were still continuing to mate with each other, and only slowly became reproductively isolated. Perhaps early humans met and mated with early chimpanzees at the interfaces where forests gradually turn into more open savanna habitats. Perhaps early humans evolved while still in the forests, taking our earliest bipedal steps there.

ANCESTRAL IMPOSTERS?

The apes we see today represent the rather sparse remnants of a once diverse array of apes that flourished in Africa, southern Europe, and Asia from 13 to 5 million years ago (Table 3.1). Today there are only five types of great apes (including us in this category), but during this earlier time there were more than twenty types of medium to large apes. From Africa alone, and from nearby areas in Eurasia that had ancient climates similar to the African sites, we now know that there were probably well over ten different types of apes (see Table 3.1). "Planet of the apes" would be a veritable description of this time. Recent discoveries suggest such ape diversity existed right up until the time when early humans first emerged.[32] A healthy number of these apes bear features in their fossil bones, teeth, and jaws that seem to indicate they are closely related to the African great ape and humans. How closely related are these fossil apes to gorillas, to chimpanzees, or to humans? No one knows for sure. A number of these ancient apes likely hit an evolutionary dead end, and are not direct ancestors to any living ape or human. Since back in that early time immediately after our evolutionary separation from chimpanzees we likely appeared extremely similar to chimpanzees—probably unlike what either species looks like today, and probably much more like one of these fossil great apes—paleoanthropologists searching for the earliest human ancestor face a pernicious problem: How do we detect the first human amongst the branches of all these fossil apes?

This problem has not stopped paleoanthropologists from making claims that fossils they have discovered represent various members of the earliest stages of the lineage leading to humans. But there are two potential snags with this process. The first is homoplasy in morphological features that, as we have seen (in chapter 1), is unreliable for unraveling the relationships among living papionin monkeys. For fossil apes claimed to be early human

Table 3.1 PLANET OF THE APES: FOSSIL APES IN AFRICA, SOUTHWEST ASIA, AND SOUTHERN EUROPE 13 TO 5 MILLION YEARS AGO

Fossil Ape	Country	Million years	Lifestyle snapshot
East Africa			
Orrorin	Kenya	6.1–5.7	Medium-size, possible biped, also climbed trees, fruit-eater, hard seed/nut eating
Ardipithecus	Ethiopia	5.6	Medium-size, possible biped, also climbed trees, general fruit-eater and omnivore
Lukeino ape	Kenya	5.9	Thin enamel on molars, possible soft food diet
Ngorora ape	Kenya	12.5	Thin enamel on molars, possible soft food diet
Nakalapithecus	Kenya	10	Large-body, hard seed/nut eating
Samburupithecus	Kenya	9.5	Large-body, fruit-eating, quadruped on ground
Chororapithecus	Ethiopia	10	Large-body, plant-eating
Central Africa			
Sahelanthropus	Chad	7	Medium-size, possible biped, heavily worn teeth, diet unknown
Southern Africa			
Otavipithecus	Namibia	13	Medium-size ape, soft fruits and leaves
West Africa			
Fossil Ape	Niger	13	Medium-size ape, slender mandible
Southern Europe			
Ouranopithecus	Greece/Turkey	10–7	Large-body, hard seed/nut eating, quadruped on ground, lived in dry open woodland
Oreopithecus	Sardinia/Tuscany (Italy)	9–7	Large body, leaf/fruit eating, climber, lived in swampy forest
Southwest Asia			
Udabnopithecus	Georgia	8.5–8	Smaller ape, lived in dry open woodland

ancestors, could their appearances be deceiving? The second problem is one of timing. We described how many new genomic dates estimated for the evolutionary separation of chimpanzees and humans range between four and six million years ago and therefore seem to suggest a more recent split than once thought. If these genome-based dates are correct, then fossils like *Orrorin* (dubbed "Millenium Man"), *Sahelanthropus* (nicknamed "Toumai") and *Ardipithecus*—fossils with geological dates that extend back to almost seven million years ago—would necessarily have predated the split between humans and chimpanzees and therefore would be too early to represent our earliest ancestors' bones.

Recently, the paleoanthropologists Bernard Wood at The George Washington University and Terry Harrison at New York University reviewed three main features used to identify fossils claimed to be early members of the human lineage.[33] The first is reduction in the size and sharpness of the upper canine teeth, features believed to be the start of a trend in the human lineage to small and similar-sized canines in males and females. The second feature is a forward position of the foramen magnum (the hole on the base of the skull where the vertebral column meets the skull), usually associated with balancing the head on a vertically oriented, upright vertebral column. Animals that are typically quadrupedal have a foramen magnum placed towards the back of their skulls. Third, there is the collection of features of the hip, thigh, leg, and foot associated with upright walking in humans.

But how exclusive are these three features to the human lineage? Paleoanthropologists have often looked at fossil sequences from past to present linking up fossils with living primates, but Wood and Harrison took a fresh perspective by looking sideways at the fossil record, as if they took a time travel machine into the past, looked around like ancient naturalists, and took note of the wide diversity of apes from this ancient time period from Europe, Asia, and Africa. Interestingly, the apparently distinguishing features of early humans can indeed be found in one or another of these fossil apes, perhaps not all together but at least separately. This implies that assumed uniquely human features could very well be homoplastic, separately evolved in different fossil apes, and thus are unreliable as evidence to claim a particular set of fossils belongs on the human family tree. For instance, their analysis challenges the idea that features we associate with upright walking in humans are always associated with this form of locomotion and suggest that anthropocentric paleoanthropology may lead scientists to interpret human-like features of a fossil with respect to their functions in modern humans. When you look sideways at the diversity of apes at this time, you can find "bipedal" features in a number of fossil apes clearly not on the human family tree. One prime example is *Oreopithecus*, a fossil ape dating between 9 and 7 million years ago from areas of Tuscany and Sardinia in Italy. It is quite remarkable that we have a nearly skeleton of this ape. I remember *Oreopithecus* fondly since Terry Harrison was actively researching this fossil ape while I was his student in the late 1980s and early 1990s. Though for much of this time I was still an undergraduate, Terry actively bounced his ideas off me and other students as he worked on this project. We know now that *Oreopithecus* is fairly certainly not on the human family tree; however it had pelvic and thigh bones that had distinctly human-like features as well as a forwardly positioned foramen magnum. These are all features usually associated with bipedalism and thought to be unique to the human lineage, though Harrison believes them to be associated with *Oreopithecus'* ability to climb vertically up trees and maintain an upright posture once up in them. Why aren't the features indicative of a bipedal walking ancient Italian ape? Because there are a whole number of features of the shoulders, arms, and hands as well as feet that indicate *Oreopithecus* was a tree climber. Surprisingly, *Oreopithecus* also happens to show small and less sharpened canine teeth, another feature thought to be found uniquely in human ancestors.

In short, features believed to be uniquely human may in fact not be, and may fool us into thinking a particular fossil is an ancient member of the human family tree when it is not. Features of the skeleton we might think are associated only with upright walking might in fact have evolved in certain extinct apes because they play important functional roles in locomotor styles other than bipedal walking. Although it is possible that some of the fossils mentioned above are members of the early human lineage, these lessons should make us cautious in our interpretation of ape fossils from around the time of the human-chimpanzee split. We cannot simply assume that certain human-like features necessarily indicate the fossil is an early member of the human evolutionary lineage.

Fossils claimed to be from our earliest ancestors also must fit into a time frame that is consistent with genome estimates of our split from chimpanzees. While many estimates from genomes at present indicate a human-chimpanzee split happening in the time-frame of around four to six million years ago[17,18,30,34,35] this is a time too recent to comfortably accommodate *Ardipithecus, Orrorin,* and *Sahelanthropus* as our earliest human ancestors. It is possible, however, that our estimates may change as we obtain a better understanding of the rate at which DNA changes accumulate in the genome. For example, some researchers have estimated that the mutation rate that produces DNA differences between humans and chimpanzees may have been slower than we have thought. If the DNA differences between the two species accumulated at a slower rate, this would actually push the split between humans and chimpanzees deeper into the past,[15] and would give us a better idea of when we can expect to first see early humans in the fossil record.

The realization that a recent separation between the chimpanzees and humans, between four and six million years ago, would conflict with the early ages of well-known fossils has spurred some geneticists to put forward possible explanations. The evolutionary geneticist Arndt von Haeseler and colleagues at the Max F. Perutz Laboratories in Vienna propose that ancient gene trees could provide a partial explanation.[36] Remember that for about one-third of our genome, we are not most closely related to chimpanzees (and bonobos). This part of our genome is very ancient and emerged even prior to the time when gorillas split from chimpanzees and humans. Let's look at the gene tree inside the great ape and human species tree that represents such an ancient region of the human genome (Figure 3.8). The tree on the left shows that the gene tree (the skinny tree) does not agree with the species tree (the fat tree), since as we proceed backward in time from the present, chimpanzee and gorilla gene copies coalesce more recently in time than either one of these ape gene copies coalesces with the human gene copy. In this illustration, the human gene tree branch is longer (and more ancient) than the gene tree branches leading to either of the apes. On this ancient human branch, DNA changes (represented by stars) evolved at a time well before humans and chimpanzees split in the species tree. Importantly, these DNA changes did not evolve on either of the ape gene branches and therefore today are uniquely found in humans. Such DNA changes could represent the earliest of all human DNA differences from the apes. (For much of our genome, though, we believe that uniquely human DNA changes evolved much more recently in time, as seen in the species tree on the right in Figure 3.8.) Nevertheless, von Haeseler suggests that ancient DNA changes could be the genetic underpinnings of some of our unique features such as skeletal traits related to bipedal walking or human-like reduced canine tooth size. If this hypothesis is true—and it is far from being tested as yet—it would mean that some of our unique human features evolved several millions of years before the human lineage evolved! How can this be? Well, it might help if you imagine this ancient ancestral species—before gorillas split off the species tree—distributed throughout large regions in Africa and also being quite genetically

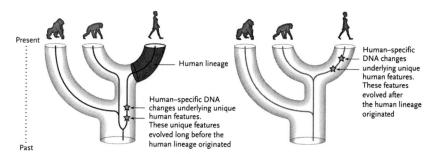

Figure 3.8: On the left, the gene copy from humans coalesces with the copies from chimpanzees and gorillas very deeply in the past before the gorilla branched from the species tree. In this scenario, unique DNA changes (stars) in the human copy of the gene could be extremely ancient, evolving well before the human lineage evolved (dark gray shading). On the right, gene copies between chimpanzees and humans coalesce very recently in the past. For this gene tree, unique DNA changes in the human copy of this gene are much more recent and confined to the human species' branch.

and anatomically diverse. A widely spread species might very well have consisted of various different subpopulations in which slightly different features existed. Perhaps in one of these ancient subpopulations, human features were already evolving.

How does von Haeseler's hypothesis provide a possible explanation for how unique human features could have wound up in early fossils, fossils that may be too early to actually represent the bones of our ancestors? In Figure 3.9, a species branch leading to *Sahelanthropus tchadensis*, one of the purported early human fossils dated to around seven million years, is placed in such a position that indicates it does not represent a human ancestor. Notably, the gene tree branch within the *Sahelanthropus* species coalesces with the human gene branch before coalescing with any ape gene copy. For this specific gene, then, *Sahelanthropus'* gene copy was more genetically similar to the human gene copy than to the gene copies in either the chimpanzee or gorilla. Therefore, for this specific gene, *Sahelanthropus* had the same unique DNA changes (denoted by stars) that humans had. If these DNA differences were responsible for at least some uniquely human features, then *Sahelanthropus* would have had them too. But for other genes, *Sahelanthropus* might in fact have been more genetically similar to the other apes, and this mixture of genes could have led this extinct species to have a mosaic of ape and human features.

For now, von Haeseler's hypothesis is a possible explanation for why some early fossils have certain unique human features even though they are not truly human ancestors. But there are many questions that need

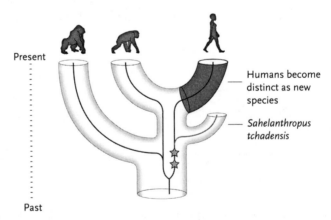

Figure 3.9: The same species tree is shown here as in Figure 3.8, but with an additional branch to the extinct fossil species known as *Sahelanthropus tchadensis*, dated to between six and seven million years ago. It is quite possible that this early fossil species (claimed by some to be the earliest human ancestor) inherited some of its human-like features because it inherited DNA changes (two stars) that evolved on an ancient human gene lineage well before the human lineage ever evolved.

further investigation. One of the most important questions is: Do ancient genes within our genome—those genes that coalesce deep in the past—actually have special DNA changes that code for unique human features? Our understanding of the links between specific DNA changes in our genome and specific anatomical features is still in its infancy. With complex features like skeletal traits, we expect there to be many DNA changes responsible for them, and we expect that these DNA changes will not have occurred only in a single gene but in numerous different genes. Until we sort out these questions, we cannot know if ancient human genes can explain our contentious "earliest ancestors." But the ancient human genes hypothesis represents a fundamentally different explanation from that of Wood and Harrison, where human-like features in apes are the result of homoplasy—similar adaptations evolving independently and having a different genetic basis. In von Haeseler's explanation, the features are genetically identical. Nonetheless, both explanations—ancient human genes and homoplasy—might work together to provide a meaningful explanation for different early fossils that contain human features even though they might not be on the human lineage at all.

As we have seen, one of the biggest reasons why many human genes coalesce deeply in the past, thereby leading to the confusion of who is our closest ape relative, is because the size of the population of our most recent common ancestor shared with chimpanzees was relatively very large. In the next chapter, we explore how a reduction in population size along the human lineage has altered the trajectory of our evolution in sometimes negative ways.

CHAPTER 4

A Population Crash in the Past

In his 2013 book, *The Story of the Human Body: Evolution, Health and Disease*, the anthropologist Daniel Lieberman at Harvard University, explains that our species may be maladapted in numerous ways as a result of our evolutionary past.[1] Here in fact, our uniquely small effective population size over the millions of years of our evolution may have been a contributing factor. There are some unexplained but tantalizing clinical findings that may indicate that this is indeed the case. Chimpanzees, our closest cousins, do not suffer acquired immune deficiency as a result of HIV infection, nor a variety of other very serious maladies that we face. Many of the complex diseases prevalent today—type 2 diabetes, Crohn's disease, rheumatoid arthritis, asthma, and others—seem to be caused by damaging yet rare DNA variants spread throughout our genome. Could it be that our unique evolutionary past allowed such damaging DNA variants to slip through the cracks to later emerge as the root of the chronic diseases we face today?

In the previous chapter we saw that the effective population size of humans is consistently estimated in genetic studies to be approximately 10,000. We also saw that estimates of the effective population size of the ancestral population that eventually split to form the human and chimpanzee lineages was relatively large, around 50,000 to 100,000. In fact, going further back in time, the effective sizes of the ancestral populations we shared with gorillas and orangutans were also relatively large. Comparing these numbers between the apes and humans indicates that a drastic reduction in effective population size must have occurred as humans evolved out of their ancestral roots among the apes, a nearly fivefold to tenfold reduction.

In the face of this "population crash," what can new genomic evidence tell us about the evolutionary consequences of this population reduction during human evolution? How might the dramatic crash have impacted the trajectory of human evolution? Has a population reduction taken a toll on the health of humans today? On the flip side, were there any possible benefits we have enjoyed as a result of living in reduced populations? And what will our vastly larger population numbers on this planet hold in store for us in our evolutionary future?

WHY SO SMALL?

How do geneticists estimate ancient population sizes? One of the most common ways is through measuring amounts of DNA differences among the genomes of individuals living today. As a general rule, the more DNA differences (or genetic variation) we see among the individuals in a population, the higher the estimate of the effective population size in ancient times and vice versa. This is because the DNA differences we see today arose in the people living in ancient populations hundreds of thousands to even a few million years ago. The process of mutation is relatively slow but ongoing, and DNA differences are created between and among people relatively constantly over evolutionary time.

The more people in a population, the more DNA differences will occur among people within the population. The fewer the number of people, the fewer opportunities for the mutational process to create DNA differences among people. Think of shooting targets in a carnival game with a friend, and neither one of you is a really good shot. Your friend has an obsession for shooting the little paper bullseye targets (and also for winning the giant stuffed panda); you are far less enthusiastic. While you take five turns shooting at five different paper targets (only one shot per target), your friend takes fifty turns shooting at fifty different targets. He would have fifty pierced targets, while you would only have five. But remember, you're both equally bad shots. While you may both have bulletholes scattered widely over the area of your targets, he will have many more. Some targets will have holes way out on the margin of the paper, others somewhere in the middle, and others closer to the center. Similarly, in a large-sized population there are more mutational targets (more people) and therefore more opportunities for the mutational process to produce DNA differences among people. In a smaller population, there are fewer mutational targets (fewer people in the population) and as a result fewer opportunities for the mutational process to produce DNA differences.

There is another very important factor that explains why DNA differences among people are fewer in smaller populations compared to larger populations. The evolutionary force known as random genetic drift is the responsible factor. This random evolutionary force always acts to eliminate DNA differences in populations, except its effects are always more severe in smaller as opposed to larger populations for reasons that we will describe below.

Some of our earliest knowledge about the DNA differences between people around the world came from studies in the late 1980s and early 1990s by the late Allan Wilson, and his students at the time, Mark Stoneking, Rebecca Cann, and Linda Vigilant. Our mitochondrial genome is vastly smaller than the nuclear genome contained within the nucleus of our cells.[2,3] One finding that gained attention from these early mitochondrial DNA studies, for instance, was that African peoples hold higher levels of genetic variation than Europeans or Asians (a finding confirmed in more recent studies of the larger nuclear genome). This was interpreted as indicating that Africans represent the oldest human populations and that Africa is the geographic origin of our species. One of the most surprising findings, however, was how very little mitochondrial DNA variation there was overall in the human population, with the effective population size of humans estimated in the range of 5,000 to 10,000.

While the mitochondrial DNA studies represented a milestone in our knowledge of our evolutionary origins, the results from these studies needed to be put into perspective. The mtDNA genome represents only a very tiny fraction of our entire genome, less than one-thousandth of 1% of the nuclear genome. The mitochondrial genome also has a completely different mode of inheritance, being inherited only through the maternal line, whereas the nuclear genome (except for the small Y chromosome) is inherited through both parents. The nuclear genome has thousands of independently recombining units each of which can serve as separate and independent test of the early mitochondrial DNA findings. Thus, it was imperative to examine the much larger nuclear genome inherited though both parents.

In order to compare the mtDNA findings with the nuclear genome, in 1999 Jody Hey and I sequenced a portion of a gene known as *PDHA1*, which codes for an enzyme in energy metabolism, on the X chromosome in a worldwide sample of Asians, Europeans, and Africans. While *PDHA1* was consistent with the mitochondrial evidence in indicating that the origins of the anatomically modern human species was in Africa, it did offer some very interesting new insights. Using a molecular clock, we traced the DNA differences in this gene among people today back to very ancient times, well before the 200,000 years represented in the mitochondrial genes. We were able to trace DNA variation in *PDHA1* back to almost two million years ago, a time when *Homo habilis* lived,[4] establishing one of the oldest evolutionary gene trees for humans at that time. Although, a published commentary on our work referred to the results as the "X files," one very important discovery we made was that some of the DNA differences between modern people today stem from very ancient times indeed.

In the years following our research on the *PDHA1* gene, researchers determined the DNA sequences of several more nuclear genes from individuals from diverse world populations, and found genetic variation extending back to almost two million years ago.[5,6] The surprising results of the "X files" no longer seemed so surprising, and a new perspective on human evolution was emerging. While each nuclear gene yielded slightly different estimates of effective population size, most estimates centered roughly around 10,000.[7] The largest study to date, in which subsets of DNA regions were sequenced in large numbers of people, analyzed forty separate regions from different chromosomal locations of the nuclear genome in people from around the world. A considerable number of the gene trees from these forty regions are ancient, extending back millions of years into the human evolutionary past. When the researchers assessed the genes as a group, they found the DNA variation in these genes was compatible with an effective population size of around 14,000.[8]

Full genomic comparisons among people from around the world are now possible. Adam Siepel's group at Cornell University has used a novel method to estimate effective population size for humans. Instead of studying the genetic differences in genes from many people around the world, they've analyzed the entire genomes from six diverse people from different world populations: a Korean, a Chinese, a European, a Yoruban (a large West African ethnic group), a Bantu African, and a Bushman from southern Africa (speakers of the click language known as Khoisan). The effective population size estimate from this study was nearly 9,000, quite consistent with the 10,000 figure.[9] It may be surprising that such a small number of genomes could contain information about past population size, but this finding directly stems from the fact that each genome is made up of many independent segments, all of which have been shuffled over the long course of human evolution due to the process of recombination. Therefore, even one full genome, with its many segments having been sequenced, can be used in a way analogous to studying one gene segment that has been sequenced in many people.

Because the gene trees built for these "ancient" nuclear genes have ranged from over one million years to even three million years, we can see that human populations seem to have been relatively small (in terms of effective population size) over a very long period of time of our evolution. However, the fossilized remains of human ancestors indicate that they were sometimes spread over the entire Old World; for example, *Homo erectus* in Africa and Asia. To occupy such vast geographic regions their population sizes surely must have been much, much larger than an effective population size of only 10,000 (i.e., a census size roughly estimated at 30,000) would indicate. But, while the census size of past hominins must have been much larger than 30,000, it is important to remember that effective population size is a measure *only* of the number of individuals in ancient populations that interbred and contributed genes to modern humans today.

WHAT ARE THE GENETIC CONSEQUENCES OF SMALL EFFECTIVE POPULATION SIZE?

Thinking about the genetic consequences of small effective population size over such a long evolutionary time period poses a challenge, one that we can better understand by looking at the consequences of population size reduction in human populations of today, in populations that we call isolates.[10] The Amish, the Dunkers, and the Hutterites are distinctive religious populations in North America, and there are similarly small isolated populations around the world such as the Paisa of Colombia. Many also adhere to cultural practices that prohibit marrying outside the group. It is crucial to understand that both these demographic occurrences have profound consequences on the genetics of the population. When these populations were established, the genetic variation carried by the few original founders was merely a small and random sample of the much larger set of genes found within the populations from which they came. For example, the Amish represent a subgroup of Mennonites who left Switzerland in the early 1700s and settled in the state of Pennsylvania in the United States.

When a small group of founders starts a new population, only a small and *random* sample of genes exists in the new population. This sample is very different in its diversity from the genetic diversity within the larger population from which it came. We can think about the genetic consequences of such a founding with another analogy. Suppose you took a handful of marbles from a bucket containing 100 marbles (these are the genes in the original large Mennonite population) in which there are ten types of differently colored marble (different versions of the gene) thoroughly mixed up inside the bucket. How many different colors would be represented in your hand? You may have retrieved perhaps three, four, or five marbles of different colors, but it is very unlikely that you would have a handful containing all ten colors. This loss of genetic variation is known as a *founder effect*. Even after many generations, the population—if culturally isolated from surrounding populations—will still have a restricted and limited set of genes due to the initial founding event.

The fact that these populations have remained on the small side and/or do not marry outside their cultural group also intensifies random evolutionary processes. To examine these random effects, let's take the simple example of flipping a coin heads or tails. Suppose that heads and tails represent two versions of a gene, with the A version (or allele), coding for the A antigen on people's red blood cells and the B allele coding for the B antigen on blood cells. We'll start out with a population in which half the versions of genes in the population are A alleles and the other half are B alleles. If we flip the coin ten times (akin to the mating among a small group of people), how likely will it be to get a 50:50 ratio of heads to tails (or five A alleles and five B alleles) at the end of ten flips? The chance is relatively small, around 25%. There is a larger chance, about 75%, of not getting five heads and five tails but getting either four heads and six tails or three heads and seven tails—or getting either all heads or all tails. When mating occurs within small populations, random processes are augmented. There is a relatively large chance (compared to mating in a large population) that the proportion of one allele decreases while the alternative allele increases, or even that one allele is lost completely, in which case the alternative allele becomes the only allele in the population.

Now, let's assume we flip the coin one million times! In a large population, there is a much greater probability of maintaining both the A and B alleles at or close to the initial 50:50 ratio. Complementarily, the chances are very small of getting a ratio that deviates very much at all from a 50:50 ratio. (This is because of a basic principle in probability: the more events you have, the more the events will tend to a 50:50 distribution, and the less you expect deviation from a ratio of 50:50.) When mating occurs in larger populations, there is a relatively decreased probability (compared to mating in a small population) of deviations in the proportions of alleles over time, and it will be less likely that any single allele is lost.

These two components we have described—founder effect followed by small population size—are both forms of an evolutionary process known as *random genetic drift* (in shorthand, just "drift") which tends to lead to populations with reduced genetic variation over time. Certain DNA variants, even though they may be harmful, may become common in small populations solely through the action of drift. One such extreme example has been found among the native people on Pingelap Atoll, one of the many islands in Micronesia, made famous by Oliver Sacks in his book, *The Island of the Colorblind and Cycad Island*.[11] In 1775 the Typhoon Lengkieki ravaged the island, leaving only twenty survivors. One of the survivors was a Pingelapese ruler, Nahnmwarki Mwahuele, who was thought to have been a carrier for the disease achromatopsia, which causes total colorblindness due to a lack of functional cone cells in the retina. The ruler's genetic contribution to the initial generations following the typhoon, through his many surviving children, was evidentially large. This resulted in a founder effect whereby the allele for achromatopsia became disproportionately common in future generations. Due to inbreeding amongst the island's small population, random drift caused the deleterious allele to become increasingly common until today nearly 10% of the population is afflicted with total colorblindness, with an additional 30% of the population acting as carriers of the achromatopsia gene. In the world population, by comparison, achromatopsia is rare and only found in about one in 50,000 people.[12] Other small island populations offer good examples of the consequences of founder effect and drift. The population of Tristan da Cunha, located midway between Africa and South America (and the only inhabitable island found in this region) shows an increased prevalence of asthma, up to 47%, among inhabitants of this island,[13,14] whereas in the United States the prevalence is much lower, approximately 7%.[15]

The same types of random evolutionary processes have affected other isolated populations established by small founder groups, such as the Old Order Amish, Dunkers, and Hutterites, populations in which there is an overall loss in diversity in DNA variants compared to the European populations they came from. Certain alleles have become more common merely through random drift acting over generations in these small populations. For example, in the 1950s the famous American geneticist Bentley Glass studied genetic drift in the Dunkers,[16] a German Baptist group from Pennsylvania named for their special practices in which a person is baptized three times, once for each member of the Holy Trinity. In this group Glass found a marked increase in Type A blood (60%) compared to the levels of Type A in the German population from which the Dunkers were derived (40%), and a much reduced frequency of Type B blood. The Dunkers are also more likely to have unattached earlobes (rather than attached) and less likely to have "hitchhiker's" thumbs—a thumb that bends way back (rather than not)—compared to other populations. These traits are common variations that geneticists have long studied in world populations, and it was easy for Glass to compare the prevalence of these traits in the

Dunkers with their prevalence in other larger populations. In other small isolates we find similar increased shifts in trait frequencies due to drift. Some traits have no serious biological significance, as with the earlobes and hitchhiking thumbs. But there are serious deleterious traits and genetic disorders that are found at higher rates in these small populations, compared to the general population. For example, Bowen-Conradi syndrome (a developmental and neurological disorder) afflicts one in 355 births in the Hutterites but only a few cases are known worldwide;[17] Ellis-van Creveld syndrome (dwarfism and polydactyly with heart and respiratory complications) occurs in one of 5,000 births in the Amish but in one in 60,000 to 200,000 births in other populations;[18] Tay-Sachs disease affects nearly one in every 3,500 births among the Ashkenazi Jews, about one hundred times more frequent than in the general population;[19] and in Native American groups from the southwestern United States, albinism appears in one in 140 in the Jemez tribe, compared to one in 3,750 in the Navajo tribe, yet it affects roughly 1 in 36,000 of European descendants in the United States.[20]

AN EVOLUTIONARY SEESAW

Of course there is another evolutionary force that operates simultaneously alongside random genetic drift. This force is natural selection, the powerful mechanism of evolution outlined by Charles Darwin in *The Origin of Species*. Natural selection, however, is different than random genetic drift because it is nonrandom, and is often referred to as a deterministic force. If a new trait that is beneficial enters a population, there is a good chance it will spread through the population by the process of natural selection because the trait increases the fitness (survival and reproductive output) of individuals who possess it. The two evolutionary forces—genetic drift and natural selection—differ in that one is random and the other nonrandom, but both forces are always acting simultaneously and both jointly determine patterns of variation within a population. We can imagine the evolutionary process as a "seesaw," with natural selection sitting on one side and genetic drift sitting on the other (Figure 4.1). At any one time, one side of the seesaw may be heavier than the other side. In other words, at any given time either natural selection or genetic drift will be dominating over the other force within a population.

As we have seen, natural selection can promote the spread of a beneficial trait within a population. However, it can also act to remove a harmful trait from a population because individuals unlucky enough to have inherited the harmful trait have relatively reduced fitness compared to those without

Figure 4.1: The balance of evolutionary forces that act on DNA variants in a population.

the harmful trait. While the first type of natural selection is known as "positive selection," the second type is called "negative selection." Negative selection is also known as purifying selection, since the process actually helps to remove harmful traits from the population.

So, why hasn't negative selection removed the harmful traits in the Pingelapese, the Amish, the Hutterites, and other small population isolates? The answer lies in understanding how drift and natural selection interact in influencing the genetics of populations. Drift and natural selection (both positive and negative) share an inverse relationship with one another. When drift becomes stronger, natural selection becomes less effective and vice versa. We have already seen that drift becomes increasingly influential in determining the fate of alleles as populations become smaller. In contrast, the effectiveness of natural selection declines in smaller populations. As population size increases, random genetic drift becomes weaker and natural selection becomes an increasingly effective force in determining the fate of alleles— for instance, whether a beneficial allele spreads to more and more people, or whether a harmful allele is purged from the population (Figure 4.2).

Now it becomes clearer why negative selection has limited power to remove harmful traits from small isolate populations. Basically, negative selection just doesn't have the strength to do the job because it has been overwhelmed by more powerful random forces holding sway over the harmful traits in small populations. If the populations had not been founded by such small numbers of colonizers and/or had grown to be larger in size over time, negative selection would become more effective at preventing harmful traits from becoming widespread. In fact, the harmful traits that are relatively common in the small population isolates are all at much lower prevalence in larger populations because negative selection is much stronger in these larger populations and can more effectively weed out the harmful DNA variants underlying these traits.

Positive selection, too, has the same inverse relationship with random genetic drift. In smaller populations, the effectiveness of positive

Smaller Population

Larger Population

Figure 4.2: The balance of evolutionary forces within a population is greatly influenced by the effective size of the population. As populations become smaller in size (on the left of the figure) the role of random genetic drift becomes relatively greater and the effectiveness of natural selection becomes reduced. As populations become larger (on the right), just the opposite is the case.

selection—its power to spread beneficial DNA variants throughout the population—becomes less effective since random genetic drift dominates. In large populations, however random genetic drift loses its power and positive selection's power increases. What does this mean, then, for a new beneficial trait that arises in a population? How does this trait fare in populations of different sizes?

Suppose a beneficial trait arises within a large population that increases the survival and reproductive success of individuals who bear the new trait. What will happen to this trait? If it isn't lost immediately after it arises (after all, it is present initially in only a single individual and could easily be lost right away by chance) there is a good probability it will spread over generations as more and more offspring are born to parents who possess the trait. This is even true for traits that have only very slightly beneficial effects, but in those cases the population size must be quite large. It is a different story in small populations, where the spread of the beneficial trait is much reduced because random genetic drift dominates over positive selection. One qualification is necessary, though; it depends on exactly how beneficial the trait is. If a beneficial trait is extremely beneficial and increases one's fitness greatly, the trait will tend to spread even in small populations where random genetic forces are powerful (although this happens at a much slower pace than in a larger population).

One example where strong drift seems to have prevented the spread of a beneficial trait is on the Melanesian Islands of the South Pacific Ocean.

Recent genetic research has shown that there is strong genetic drift acting on each of the islands. Since drift acts in a random way, we would predict that the populations on the different Melanesian Islands would be genetically different from each other—just what the genetic studies show.[21] For instance, there are significant skin color differences between the islands.[22] While all Melanesians tend to be dark skinned, there are marked skin pigmentation differences between people inhabiting different islands, especially in people from the island of Bougainville, who have the darkest skin. This finding is peculiar since all the islands fall at latitudes between one and seven degrees south, which means that the sun's ultraviolet light intensity should be similar among the different islands. We might expect there to be strong selective pressure for very dark skin on all islands, and that the inhabitants of different islands should all be rather homogeneous for skin color. However, high levels of genetic drift in these small island populations have probably played an important role in limiting the effectiveness of positive selection in maintaining a more homogeneous dark pigmentation. Interesting, too, is the fact that the island most geographically isolated from the others, and therefore prone to the most drift (because of less between-island breeding), is the island of Bougainville, exactly where skin color is most different.

Now let us suppose a deleterious trait arises in a relatively large-sized population. In this situation, the power of drift is much reduced in relation to the effectiveness of negative selection. Negative selection will weed the trait out of the population, either through mortality or reduced reproductive success of individuals bearing the trait. Through negative selection, the harmful mutation will be prevented from spreading beyond more than only a small fraction of people and might even be removed entirely from the population.

In small populations, however, the probability increases that the trait is able to spread through the population, even if it is deleterious. Again, though, the degree to which the trait is harmful needs to be taken into account. If the trait is very harmful, then it is likely that negative selection will remove the mutation or keep it rare even in a very small population. If the mutation is only slightly or moderately harmful—having a small negative impact on fitness—then it is possible that the harmful trait can spread within the population and become somewhat common.

OUR ANCIENT SEESAW

We can now return to the question: What consequences were there of a fivefold to tenfold reduction in population size over millions of years along the human lineage? Can we detect similar consequences to those that we see in small isolate populations today? Since the entire lineage experienced small effective population sizes, can we see the consequences of this in our genomes?

Based on our theoretical understanding of the balance between drift and natural selection, we can make several predictions about how the human lineage would have been affected. Luckily, we are able to test these predictions because we have the full genomes of a number of organisms to use for comparison. Species like mice, rats, and fruitflies are especially useful to us because their population sizes are vastly larger than that of humans, and therefore the consequences of large population size, and its effects on the relative powers of drift and natural selection, can be seen more easily.

To consider random genetic drift first, because of considerable size reduction on the human lineage, we can predict that the seesaw should tip in the direction of random genetic drift. That is, drift should have become relatively more powerful along the human lineage than it had been in the common ancestral population we shared with chimpanzees. Thus, we expect that during human evolution a stronger random force held sway over which traits, and the genes that underlaid them, remained in the population. If random forces had become increasingly powerful, it would mean that natural selection became less effective relative to its efficacy in the larger ancestral population we shared with chimpanzees. That is, both types of natural selection—negative and positive selection—should have become weaker along the human lineage since we split from chimpanzees. But what kinds of evidence should we look for in genomes to test our predictions?

We might begin by measuring the strength of negative and positive selection along the human lineage by analyzing DNA changes in genomes. We could then compare these measurements with those taken from other genomes, particularly genomes from organisms with larger effective population sizes in which natural selection is expected to have operated more strongly or more effectively over their evolution. Such analyses have already been undertaken by comparing full genomes. For these comparisons, most studies have focused on the protein-coding portion of the genome (i.e., the genes) since we have well-developed methods designed to detect the effects of natural selection in these regions.

To understand how the methods work, we need to give a little more information on how DNA codes for proteins. As we have mentioned, proteins are made up of building blocks of amino acids. Each amino acid is coded for by three DNA bases, known as a triplet. So if a protein has 100 amino acids, the gene will consist of 300 nucleotide bases. Of the three DNA letters that code for a particular amino acid, however, in most cases only the first and

Second Letter

First Letter		T	C	A	G	
		TTT TTC Phenyl-alanine / TTA TTG Leucine	TCT TCC TCA TCG Serine	TAT TAC Tyrosine / TAA Stop codon TAG Stop codon	TGT TGC Cysteine / TGA Stop codon TGG Tryptophan	T C A G
	C	CTT CTC CTA CTG Leucine	CCT CCC CCA CCG Proline	CAT CAC Histidine / CAA CAG Glutamine	CGT CGC CGA CGG Arginine	T C A G
	A	ATT ATC ATA Isoleucine / ATG Methionine; initiation codon	ACT ACC ACA ACG Threonine	AAT AAC Asparagine / AAA AAG Lysine	AGT AGC Serine / AGA AGG Arginine	T C A G
	G	GTT GTC GTA GTG Valine	GCT GCC GCA GCG Alanine	GAT GAC Aspartic acid / GAA GAG Glutamic acid	GGT GGC GGA GGG Glycine	T C A G

Figure 4.3: The full genetic code. Each amino acid is coded for by a triplet of DNA. For example, the amino acid Valine is coded for by any of four different triplets: GTT, GTC, GTA, and GTG. For many amino acids, if the third nucleotide of the triplet mutates to another base, the triplet will still code for the same amino acid. For this reason these sites are known as silent DNA sites. On the other hand, if the first and second nucleotides mutate, then the amino acid will usually change. For this reason, these sites are called amino acid-altering DNA sites.

second of these actually dictate the particular amino acid, with the third DNA base being free to vary (see the genetic code in Figure 4.3). Therefore, DNA sites within the triplet can be divided into two types, those in which a base change would alter the amino acid (amino acid-altering sites), and those in which a base change does not alter the amino acid (silent sites). Once they are classified in this way, we have the grounds for measuring the strength of natural selection.

Over evolutionary time, negative selection tends to remove most amino acid-altering mutations. Since base changes are due to the mutational process, which is slow and relatively constant over time, almost every time a mutation causes a change at an amino acid-altering DNA site in an individual, this base change will be eliminated through death of the individual or through decreased reproductive output (few or no offspring). The reason is that most proteins play important or even vital functions in our body and any alterations to them are likely to be detrimental, either to a minor or more extreme degree. On the other hand, because base changes at silent sites will not result in any amino acid change, negative selection will not remove these mutations from a population. The result is that for the majority of our genes, the rate of base changes at altering DNA sites will be considerably lower than the rate of base changes at silent sites.

For any gene, we can now measure the degree to which negative selection has removed detrimental changes by dividing the rate of changes at altering DNA sites by the rate of changes at silent DNA sites. This measurement tells us how strongly negative selection has operated on our genome over our evolutionary past. But to fully understand how negative selection has acted in the past, this value must be compared to a reference point known as neutrality. This point is a situation in which there is an equal rate of amino acid-altering and silent mutations. This point will be one, because the numbers we are dividing are the same. The point of neutrality can also be expressed as 100% and indicates that there were no more changes at altering DNA mutations removed than at silent mutations. For most proteins, however, the measure will usually be less than 100% and typically very much less than 100%, because negative selection is quite effective at removing detrimental changes to proteins caused by any DNA base changes.

THE DOWNSIDES OF A CRASH

Now we can measure how effective negative selection is at removing deleterious changes to proteins in the evolution of any species. We can do this by applying our method across all genes in a species' genome to estimate how effective negative selection was in that species' evolutionary past. Since the genomes of a number of species are now available, we can compare their genomes and ask: Was negative selection stronger or weaker for a particular species? And was negative selection stronger or weaker in humans with a small effective population size, compared to species with larger effective population sizes?

An analysis of the genomes of various mammals (including ours) to estimate the level of negative selection in their evolutionary past was performed in 2008 by Adam Seipel at Cornell University and his colleagues.[23] The value they estimated for the fraction of amino acid-altering mutations that escape removal by negative selection is approximately 25%. But what does such a number mean? In fact, such a figure suggests that approximately 75% of the altering DNA changes (the complementary fraction) that arose during human evolution were removed at some time after they arose. Now consider an animal species with a very large effective size: the mouse. Its effective population size is around 450,000–810,000,[24] about fifty- to almost one hundred-fold larger than in humans. The fraction of amino acid-altering mutations estimated to have accrued during mouse evolution is approximately 13%. This means that 87% of all altering DNA base changes were removed by negative selection during the evolution of the mouse and indicates that negative selection during mouse evolution has indeed been much more effective at removing altering mutations than during human evolution. In fact, during the evolution of the mouse, negative selection was able to remove 12% more amino acid-altering mutations (subtracting 75% from 87%) than were removed during human evolution.

We find a similar case when we compare the human genome with the genome of the rhesus macaque, whose genome was fully determined in 2007.[25] (Humans diverged from these monkeys nearly twenty-four million years ago.) Estimates put the effective population size for the rhesus macaque at about 73,000. When this method is applied to the rhesus genome, the strength of negative selection was measured as 19%. Thus 81% of mutations causing an amino acid change were removed during the course of rhesus macaque evolution. In this case we also see that negative selection has been more effective—7% more effective—at removing altering mutations in the rhesus macaque lineage compared to the human lineage.

To understand the significance of this finding, consider that most altering mutations are deleterious. In fact, most are so deleterious and have such large harmful effects on fitness (survival or reproduction) that they are immediately removed from the population. That is why the percentage of mutations estimated to have been removed is so high for all three species—mice, rhesus monkeys, and humans. So what accounts for the differences in the strength of negative selection among these three species' lineages? Is there anything special about that category of altering DNA changes that were removed in mouse evolution but were *not* removed in human evolution—that 13% difference? In fact, there does seem to be something significant about these DNA changes. It seems that this 13% greater fraction of DNA changes that were removed in the mouse are not nearly as deleterious as the 75% that were removed common to both humans and mouse. The DNA changes in this 13% do not seem to have had comparably strong harmful effects. Because they seem to have very small negative influences on fitness, evolutionary geneticists call these DNA changes "slightly deleterious." It is just these types of DNA changes that can bypass removal by negative selection when the size of a population is reduced in size. To put it another way, in the larger mouse population negative selection is more powerful at weeding out many deleterious DNA

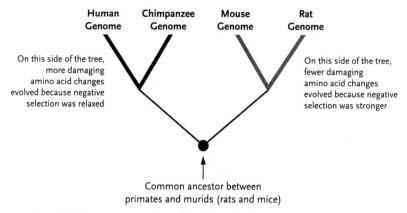

Figure 4.4: Genome comparison between two pairs of species with very different effective population sizes illustrates the balance of evolutionary forces that have shaped their evolution.

changes, even those that have only slight deleterious effects. In contrast, during human evolution negative selection was not as powerful and could not remove these slightly deleterious DNA changes. They became molded into our genomes as permanent fixtures.

There is other evidence that these mutations are slightly deleterious. When researchers closely examined the amino acids that were altered in the course of human evolution and mouse evolution (Figure 4.4), they found that amino acids within the human genome tended to have changed to more damaging types of amino acids—amino acids with very different properties—whereas in the mouse genome amino acids tended to change to amino acids that had similar chemical properties to the original type.[26] Because the switch to a damaging type of amino acid is expected to be deleterious (even if just slightly), this supports the idea that negative selection has been relatively relaxed in human evolution compared to mouse evolution.

What were the consequences of a larger fraction of slightly deleterious mutations being fixed in the human genome in our evolution? We are not sure exactly, but one possible way to find out is to look at human populations today. Remember, human populations only started to grow in size relatively recently, starting to increase slowly between 100,000 years ago and 50,000 years ago and then grow dramatically since about 10,000 years ago. For much of our species' modern evolutionary history we had a small population size. Interestingly, the modern human species is estimated to have a significant fraction of slightly deleterious mutations. We know this through analyses of large genome-wide databases that give us information about the DNA differences between the genomes of different people living today. Since alterations in protein-coding genes are of great medical interest several new studies have sequenced very large numbers of genomes of different people and reported on the amounts of DNA differences that alter amino acids. For example, in the latest efforts of the 1000 Genomes Project (available at the time of writing), the entire genomes from 1,092 individuals from fourteen different world populations were analyzed (see chapter 7 for more details).[27] Other studies have focused their sequencing efforts on just what is called the *exome*, the approximately 1.5% of the genome that codes for proteins. The rationale for this is that most important functional changes (either adaptive or disease-causing) are likely to reside in this fraction of the genome, and by concentrating on the exome it becomes more time-saving and cost-efficient to determine the sequence of even greater numbers of people.[28] These projects have shown that within the human population there are numerous amino acid-altering DNA differences between different people. But one important question was how common these types of DNA differences are, and how damaging might they be. After all, we would expect that a majority of people would not have them, since altering DNA changes would be more apt to affect a protein in negative ways. Therefore, we would predict that negative selection would "try" to keep them from spreading to large numbers of people. In fact, this is exactly what was discovered. Most amino acid-altering DNA differences are found in a small but still significant percentage of people in the population while the majority of the population has much fewer altering DNA differences. Two findings were particularly striking. About one-quarter to one-half of the altering DNA differences in human populations are likely to be slightly deleterious to a protein's function.[27] And each human individual on average carries within their genome several hundred DNA differences that alter the function of proteins, and may have negative health consequences.

Research suggests it is likely these DNA differences, simmering at low levels in the human population, contribute substantially to the burden of complex human diseases.[28,29] It seems that each slightly deleterious DNA difference does not act alone in causing complex disease. Rather, their effects may be added together, since large numbers of slightly deleterious differences are spread across the genome. And since not all people contain the same set of slightly deleterious DNA differences, each individual's degree of susceptibility to these diseases is unique. For example, in Genome Wide Association Studies (studies that search for the genetic causes of disease) thirty-two variants have been found thus far to influence the development of Crohn's disease, a type of inflammatory bowel disease. However, together these variants explain only 20% of the heritability of this disease and therefore we believe that there must be many additional variants that are responsible for the missing 80% of heritability for this disease. Likewise, it is believed that hundreds of DNA variants influence one's susceptibility to other complex diseases such as diabetes type 1, diabetes type 2, hypertension, obesity, coronary artery disease, asthma, rheumatoid arthritis, and bipolar disease, though most of these variants have not yet been found.[29] Future studies are likely to discover many more DNA variants that contribute to these complex diseases, but the discovery of these variants will only be possible once we compare the genomes of much greater sets of individuals since each variant is believed to be present in less than 5% or even less than 1% of the population. We call such diseases *polygenic*, since many DNA variants influence their expression. Since the contributing DNA differences are only slightly deleterious, each single DNA difference exerts only a small negative effect. This likely accounts for why there is a wide spectrum of severity of these diseases; some people have rather extreme cases of type 2 diabetes or hypertension while other people have only mild cases.

Returning to the millions of years over which the human lineage was evolving after separating from the chimpanzee lineage, 13% more amino acid-altering DNA changes, as we've seen, were incorporated into the human genome compared to the mouse genome, and most of these are believed to be slightly deleterious. Each one of these slightly deleterious changes might be expected to have had a slight negative impact on the fitness of our ancestors. It is possible that many of these DNA changes degraded the quality of our genome in various ways and may account for certain peculiarities of our species.

Genomics has recently detected that many unique gene losses are known to have occurred along the human lineage. These are cases where deleterious mutations disrupted the translation of the gene into a protein, resulting in the absence of the protein. Such inactivated genes are called pseudogenes and at least eighty deactivations occurred after humans separated from chimpanzees. The largest fraction of deactivated genes has had the consequence that we have lost the ability to detect an array of scents and to detect certain bitter-tasting substances—but also we have lost certain genes involved in our immune defenses.[30]

It is also truly remarkable that our closest living relatives, the chimpanzees, do not suffer from a number of our major diseases or disorders. All the following common human diseases—Alzheimer's disease, atherosclerosis, rheumatoid arthritis, asthma, endometriosis, myocardial infarction, falciparum malaria, HIV infection, epithelial cancers—are rare or absent in chimpanzees.[31] We are only at the starting point of discovering the genetic underpinnings of these huge differences in disease susceptibility but it seems quite possible that many of these differences could be explained at least partially by the increased fraction of slightly deleterious DNA changes incorporated into our genome (due to a reduction in effective population size) since we split from our most recent evolutionary cousins.

THE UPSIDE OF A CRASH

Ironically, an increase in slightly deleterious DNA changes to our genome might have also had positive consequences, and altered our evolution in positive ways. Gene losses, for example, may turn out to offer advantages. This is the so-called "less is more" hypothesis proposed in the late 1990s by Maynard Olson of the University of Washington.[32] There are several cases known in which genes losing their function (they no longer produce a protein) appears to be advantageous. For example, loss of the function of the *MYH16* gene, which codes for the protein myosin in jaw muscle attached to the side of the head, might have been beneficial to our ancestors about two million years ago.[33] Spurred by this loss of function, jaw muscles (anchored to our brain-case) would have become smaller, permitting the brain-case to become larger. Thus, the researchers suggested that the mutation released the constraint on brain size caused by large chewing muscles, and allowed brain size to increase. Another example where the loss of function of a gene might have provided an advantage is the case with the gene *CASPASE12*. In almost all humans, the *CASPASE12* gene has lost its function due to a disruptive DNA mutation. It is believed this results in a reduced susceptibility to sepsis (blood poisoning) and its lethal complications.[30] The non-functional gene is thought to have spread in human populations by positive selection. The loss of olfactory genes and bitter taste genes during human evolution presumably had the effect of reducing our sensitivity to different smells and bitter-tasting substances, but may have led to potentially advantageous forms of interaction among individuals, between humans and our environment, and the foods we eat. Much more investigation is necessary to investigate the degree to which gene loss has had positive or negative effects in human evolution, and to answer additional questions. Is gene loss a consequence of reduced population size and relaxed negative selection? To what extent is the "less is more" hypothesis true for our evolutionary past?

It is also probable that the negative effects of some or many of the slightly deleterious DNA changes to our genome were later suppressed or reversed through the effects of subsequent DNA changes that occurred later in time and at secondary locations within our genome. The interaction among two or more separate DNA changes where they have a direct functional influence on each other is called *epistasis*. Thus a slightly deleterious DNA change may be followed later in evolutionary time by a compensatory DNA change that "corrects" it. While the compensatory DNA change may often occur in the same gene affected by the original deleterious change, it may also be that the compensatory change occurs in another region of the genome. So far, it has been a difficult task to dissect out actual cases of functional interactions between DNA at different locations in the genome, and we are only at the beginning of understanding such interactions that occurred in the millions of years of human evolution. However, we do know that compensatory interactions do occur with considerable frequency in evolution. For example, studies show that about 10% of serious disease-causing mutations in humans are normally present in the genes of other species and cause no health defects.[34] It is believed that another DNA change in these species compensates or suppresses the deleterious effects of the disease-causing DNA alteration. For example, a DNA alteration in the *CRYGD* gene associated with susceptibility to cataracts (opaqueness of the lens of the eye) is commonly found in other mammals (such as the cow, dog, rat, mouse, and opossum), yet this change is always accompanied in these animals by one or two other DNA alterations located nearby in the gene that are believed to suppress the potentially deleterious change.[35] There are also a number of cases in which chimpanzee and gorilla genes contain DNA variants that cause disease in some human individuals, but are associated with normal health in these other species. The sequencing of the gorilla genome, for example, led to the discovery that a human DNA variant associated with dementia and another human DNA variant associated with enlarged heart are normally found in the gorilla and apparently have no negative health consequences.[36]

Compensatory DNA changes have also been found within the genes of some people who also carry the DNA change that normally causes disease. The compensatory changes either prevent these people from developing the disease or allow them to manifest a milder expression of the disease. Some inherited heart arrhythmias may be due to slightly deleterious DNA alterations within proteins important for normal electrical conduction in heart muscle. A number of different DNA alterations in the *SCN5A* gene are associated with rare heart arrhythmias, such as Brugada's syndrome. However, the effects of one of these damaging DNA changes can be suppressed through the alteration of a key amino acid at a different position (position 558) within the same gene.[37] Compensatory DNA changes are also known in the *ASL* gene that codes for the enzyme that normally helps individuals detoxify ammonia. Some people carry an alteration in the DNA of the gene that causes an enzyme deficiency and results in an accumulation of ammonia in the blood. Some of those people are lucky enough to also carry compensatory DNA changes in their *ASL* genes that help to restore the function of the enzyme.

Yet another way that the effects of slightly deleterious mutations may be ameliorated is through increased activity in so-called quality control genes, which are present in large numbers in our genome. Genes in this category can protect us against the effects of slightly deleterious DNA changes. For example, if a protein is assembled incorrectly because of a slightly deleterious DNA change, the proteins built by quality control genes can either help correct or degrade the bad protein so that it causes little harm in the body. (Amusingly, we call one group of quality control genes the chaperone genes.) One study of a population of bacterial cells in the laboratory forced to undergo a reduction in population-size, showed evidence of an increase in slightly deleterious mutations. (Remember, this is exactly what is believed to have occurred during human evolution.[38]) The bacteria subsequently showed hyper-accelerated evolution in their quality control genes, presumably to meet the challenge of "fixing" the defective proteins that resulted from population reduction.

Ironically, a considerable number of slightly deleterious DNA changes that arose in human evolution may eventually turn out to be advantageous. For example, researchers have long thought that beneficial DNA changes that arise in a population are beneficial from the minute they evolve. However, it is possible that many beneficial mutations may have actually started out being slightly deleterious.[39] While a slightly deleterious DNA change may have a slight negative influence on an individual's health in a particular environment, it may in fact become advantageous in a newly colonized environment, such as when anatomically modern humans spread out of Africa and into Europe, Asia, and the Americas over the past 50,000 years or so. A myriad of slightly deleterious DNA variants simmering at relatively low frequencies in the human population may have allowed some of us to quickly adapt to high-altitude environments, to cold environments, to changed diets, and changes in metabolism. Although such reversals in fortune are quite possible from a theoretical perspective, proving that a specific beneficial DNA variant today was once deleterious will be a significant challenge.

Just as with negative selection, we could expect positive selection, which promotes the spread of a beneficial DNA change throughout a population, to have decreased in effectiveness during the course of human evolution. This is especially true for DNA changes that offered only small benefits since, under conditions of reduced effective population size and the consequential increase in the power of drift, it would be difficult for these DNA changes to spread through the population. But is there genomic evidence for decreased power of positive selection during human evolution?

To answer this question, we can again examine the protein-coding portion of our genome. Our methods for estimating the degree of effectiveness of positive selection are slightly different from those we described for measuring levels of negative selection. But, as we did with negative selection, we can compare measures of the effectiveness of positive selection between species with very large effective population sizes (fruitflies and mice) and species with smaller effective population sizes, like humans. The results of these comparisons are striking. In mice and flies, a large fraction of the total amino acids that were altered during the course of their evolution—45% and higher—are believed to have been driven by positive selection because the alterations were beneficial. In contrast, during the course of human evolution, only a small fraction of amino acid alterations—around 0% to 20% with most estimates around 10% or less—are believed to have resulted from the force of positive selection. (I reviewed these results in a paper in 2010.[40]) Thus, it appears that the reduced effective population size during much of human evolution has resulted in a decrease in the effectiveness of positive selection. Presumably, the force sitting on the other side of the evolutionary seesaw, genetic drift, had gained power relative to positive selection, similar to what we saw with negative selection.

What were the possible consequences of reduced power of positive selection during the millions of years of human evolution? As with negative selection, the consequences would fall on those DNA changes in our ancestors that had small effects on fitness (survival and reproduction). In the case of positive selection, though, we are talking about mutations that offer extremely small beneficial effects, not slightly deleterious effects. Thus, DNA changes that arose in our ancestors and that had only slight beneficial effects on the fitness of our ancestors would have had a hard time spreading through the population, and maybe were lost altogether. Because these slightly beneficial mutations were either lost during our evolution, or are present in the genomes of only a small fraction of the people alive today, we have little idea about what the general consequences of relatively reduced positive selection has been on human evolution. This leaves a lot of room for speculation.

The University of Sussex evolutionary geneticist, Adam Eyre-Walker (a name that rivals Star Wars' Skywalker), has suggested several consequences of reduced positive selection in humans.[41] One possibility is that organisms with smaller populations were less able to adapt to changing environmental conditions compared to organisms with larger populations. For example, flies and mice (with larger population sizes) may have been better able to adapt to environmental change in the course of their evolution compared to humans.

Some of the ways mice and flies might have better adapted to their environments could have been through their anatomical features, but this seems odd to a paleoanthropologist. From an anatomical perspective, different species of flies or mice by and large look very similar. To us, the anatomical differences between humans from chimpanzees seem much more dramatic in comparison. But, it is possible that there are many adaptations in these other species that we are unaware of. For one, mice compared to humans are known to have powerful senses of smell as well as taste in order to detect bitter-tasting substances. Fruitflies may have many adaptations related to reproduction and physiological function that we know little about. Different species of flies have to dance in sometimes very specific ways to be found attractive by potential mates. It is also well known that different species of fruitflies are tied to very specific ecological conditions. A particular species can have very specific plant host requirements, or have specific reproductive interactions and behaviors. There are probably many biological features and physiological functions of mice and flies that we know very little about, but which are essential to their lives.

Recall, however, that the effects that differences in population size have on the force of positive selection will be greatest for mutations that have very small impacts on fitness. Therefore, if mice have been better able to adapt to their environments, it seems likely this has been only a matter of fine-tuning their adaptations through DNA changes that offered slight benefits. Even if only a matter of fine-tuning, it seems probable that through human evolution our ancestors were slightly suboptimally adapted to the environments in which they lived, and this may still be the case in our species today.

THE FUTURE OF HUMAN EVOLUTION: NATURAL SELECTION BOUNCES BACK!

Human population sizes are vastly larger today, reaching almost seven billion people with an expected world population in 2050 of approximately ten billion. Genetic information indicates that for most of the millions of years of human history we had an effective population size of 10,000 or census size of 30,000 (and perhaps larger since the conversion from one to the other is complex). It also indicates a more recent increase in human population size starting about 50,000 years ago, after our modern species began to colonize geographic territories outside of Africa. Dramatic increases in population size also began after the transitions to an agricultural and animal domestication way of life around 12,000 to 10,000 years ago and then later in the last several hundred years with the development of industrial societies.

The vastly larger number of people in recent history has almost certainly already altered the balance of the evolutionary seesaw on which natural selection and random genetic drift occupy opposite seats. In fact, the side of the seesaw where drift sits has probably already started to rise in relation to natural selection (becoming less influential), and will continue to rise. This will have the consequence of increasing the power of natural selection—both negative selection and positive selection. In effect, natural selection will rebound from where it has been over the vast majority of human evolution. Since the increases in population size happened in a micro-fraction (less than 1%) of the total time of human evolutionary history since diverging from chimpanzees, the effects of natural selection's rebound will mostly be felt by our species in the future.

We can also expect that negative selection will be much more effective in the future at weeding out DNA differences that are deleterious, even if they are only ever so slightly deleterious—although it is certainly a possibility that some of these may turn out to be beneficial in the changed environmental conditions of the future. Remember, under the reduced population sizes of most of human evolution when drift was relatively influential, slightly deleterious mutations could become quite common in the population and could even spread to all individuals of our ancestors. In the future, it is possible that the human species will be alleviated of some of the chronic diseases and other genetic quirks that now afflict some members of our species, simply because negative selection will be better able to remove the deleterious changes that are their root cause. On the other hand, advances in healthcare (e.g., medications) can greatly compensate for some negative health consequences of deleterious mutations. Positive selection will also increase in its relative effectiveness due to the dramatic increases in population size. Even DNA variants that only have very slight beneficial effects will have a higher chance of being spread by positive selection throughout populations of our species. Another consequence of the increased sizes of human populations is simply the fact that there are more targets for new mutations. The more people there are in the world, the more genomes there are. Since each genome has a certain probability of accruing DNA mutations, the more genomes that exist, the higher the number of mutations overall in the human species. Thus, there will be increased chances that a beneficial mutation will arise in at least one of the many genomes of the population. It then becomes more probable that the beneficial mutation will spread to other members of the population, ultimately increasing the overall genetic fitness of human populations. It should be noted, though, that many new mutations that arise in individuals are frequently lost soon after they arise, simply due to bad luck. On the upside, with humans numbering in the billions today, it is highly likely this same beneficial mutation will arise in multiple different people and will arise again and again in new generations.

People around the world have been having children with those from different countries or continents for several centuries, with this process intensifying, and that increases opportunities for an advantageous mutation to spread to other populations. For example, a mutation may arise in the genome of an individual in one environment where it has very little beneficial effects. However, this mutation may spread through intercontinental travel and interbreeding to individuals residing in completely different environmental conditions. In this new environmental context, the same mutation might prove to be extremely advantageous to the individuals who have inherited it from the world traveler.

The effects of massively augmented population sizes could very well lead to an evolutionary tune-up, in which diverse populations become more exquisitely matched to the varied environments in which they live. It could also lead to the honing by natural selection of our biological defenses against a myriad of pathogens from bacteria, viruses, and parasites with which we are continuously engaged in evolutionary arms races. Indeed, our species may have already started to reap the rewards of increased population size. Some researchers see evidence that human adaptation has accelerated over the past 40,000 years,[42] shortly after the time our dramatic population growth is thought to have begun. In the next chapter, however, we will take a step back and discuss how researchers have begun searching our genome for the genetic changes that have occurred over the millions of years since our split from chimpanzees and that have made us human.

CHAPTER 5

What Can the Genome Tell Us about Being Human?

Ever since Darwin, anthropologists have hypothesized about the features that set us apart from our primate cousins—bipedal walking, increase in brain size, use of spoken language, increased manual dexterity, loss of body hair, and other features. How did these features evolve? When did they evolve? Why did they evolve? Of course, all of these features are thought to have provided distinct advantages to our ancestors and to have spread through their populations by natural selection, the mechanism of evolution that Darwin proposed in *On the Origin of Species*. Because these features most probably evolved by natural selection, we call them *adaptations*. By calling them "species-wide" adaptations, I am drawing attention to the fact that the features evolved specifically on the human lineage since the time we diverged from chimpanzees (and bonobos) but before the times when our modern anatomical species differentiated into varied geographic populations with their own special features. In short, "species-wide" adaptations are possessed by all of us.

The gradual evolution of some of these features can be traced in the fossil record of our immediate ancestors because they can be identified in hard tissue remains. The transformations of these features are discernible over evolutionary time. (Of course, language and loss of body hair are exceptions.) Even though we can see some of these adaptations in the fossils of our ancestors, we have rather little idea of how our adaptations became coded within our genomes. For example, with regard to our brains, the crania of our ancestors can indicate expansion in brain size, and comparisons of the brain tissues of living humans with our ape cousins can tell us about anatomical and physiological differences between us. What fossils and brains cannot tell us are how many genes and which genes were involved, and the nature of the changes to the genome that produced our large brains and complex minds. So, how can genomes be used to investigate Darwin's conjecture of the biological continuity of the human mind with the mind of primates and with other animals?

For many adaptations, we cannot even obtain clues from fossils. The loss of body hair played an important role in our evolution, allowing early ancestors to adjust their body temperatures efficiently and to gather and hunt widely for food. Spoken language was fundamental to our complex social interactions and emotions, to communicate how to knap sophisticated burins and blades from stone, to produce a primitive flute from the wing bone of a griffon vulture, or manufacture shell and ivory beads and pendants to adorn our clothing. Yet we do not know whether these features evolved gradually or more abruptly. For these and other "un-fossilizable" adaptations, can we learn anything about them from the genome? In which ways was our genome modified to produce our adaptations, and when did these changes occur?

As you can see, there are a lot of unanswered questions about our adaptations. Fortunately now, for the first time, full genomes of humans and our closest relatives hold many clues to our adaptations and some lingering questions about our ancient past are becoming answerable. It is not exaggerating for me to tell you that exciting discoveries about the genetic basis of our adaptations appear monthly in scientific journals and that we expect exciting discoveries will continue as more genomic data and methods of scrutinizing this data are improved.

We know that brain size, bipedalism, enhanced hand dexterity, language, and loss of body hair are some major adaptations that make humans different. But if we knew more about how each of these adaptations are reflected in our genomes, we'd have a powerful tool to know how abruptly they evolved, perhaps under what kinds of conditions they evolved, and what relationship they might have to other changes in the genome. This information could be used to help us better understand some of the most controversial, most cherished, and most intriguing parts of our human "story," such as the evolution of our brain size and language and their relation to culture and tool use. Beyond that, the genome could tell us that there are other things beyond brain and hair and language that make us human, things that we might have never considered. One exciting type of analysis that can reveal such "hidden" adaptations are studies that interrogate the genome as a whole, known as genome-scanning studies. Having complete genomes is a major source of information, so we can now uncover secrets about the unknown features that make us human.

SCANNING THE GENOME FOR OUR ADAPTATIONS

We believe that evolutionary changes that occurred in our genome underlie most of our species' biological adaptations. But how to detect such regions in a genome consisting of over three billion nucleotides? And how to detect those adaptations that evolved over the millions of years since our separation from chimpanzees but before we differentiated into our many different geographic populations of today? One important way to get at our species-wide adaptations is to compare the genomes of humans with those of chimpanzees, to pinpoint locations where they show DNA differences. Humans and chimpanzees differ at about one out of every one hundred DNA nucleotides, making us only 1% different. However, this amounts to thirty million DNA differences, a heck of a lot to sort through to try to pinpoint only those differences that accumulated in humans because they were adaptive. Evolutionary theory has determined that there are two very different processes (discussed in chapter 4) that produced the genome differences between two close species. One is random genetic drift in which DNA differences accumulate at a "ticking-clock-like" rate but have no functional effect at the time they evolve. The other is positive selection, which promotes differences because they had a functional effect that was indeed beneficial. Since both act simultaneously during evolution, the trick is to decipher which DNA differences evolved because they were beneficial or adaptive, and which were simply randomly evolved differences of no functional effect.

One way to narrow the search is to concentrate our scans of the differences between humans and chimpanzees at only those locations in our genome known to code for proteins, which is approximately only 1.5% of our genome, yet contains approximately 21,000 genes. Since the first scan of the human genome was completed in 2003, numerous scans have been carried out, most searching within the protein-coding portion of our genome.

Even though genes only account for a very small amount of our genome, this amounts to searching among about 60,000 functional differences between humans and chimpanzees. Two major questions are what proportion of these differences were beneficial and enabled our human ancestors to adapt to their changing environments and evolve their special adaptations, and in which genes do these differences lie?

The signature of adaptation we look for in genes has its basis in the specific way that genes code for proteins. As mentioned in the previous chapter, DNA sites can be categorized into two types: those DNA sites where, if changed, an amino acid will alter within a protein, and those DNA sites that can freely change without causing an amino acid alteration. We called these two types of sites amino acid-altering sites and silent sites. Thus scans of the genome usually focus on genes showing accelerated change at amino acid-altering DNA sites over evolutionary time. If there is no adaptation at a gene, we expect that amino acid-altering DNA sites will have changed much less often than silent DNA sites. In fact, we predict this will be the case for most genes. After all, for most species proteins play the same or very similar roles, and the evolutionary process will act to preserve their function. But if a gene has experienced positive selection, we expect that changes at amino acid-altering sites will have occurred more frequently over evolutionary time than at silent sites (Figure 5.1). For humans, such signatures of positive selection are rare among our genes but are extremely important for us to identify so that we can better understand how we evolved our unique features.

When the results from scans for evidence of positive selection in our genome started coming in, there was great anticipation that scans would reveal the genomic locations that underlie what makes us human, all those adaptations we mentioned earlier. Sure enough, the scans turned up hundreds of candidate genes that potentially underlie human adaptations. Unfortunately, results did not match up to our most ambitious hopes. That is, the candidate genes did not seem, at least in any direct way, to explain the human brain, bipedalism, hand dexterity, language, loss of body hair, or sweating—the biggies of human evolution. Instead, large groups of these genes were related to immune function and host defense (both related to disease resistance), chemosensation and olfaction (like the senses of taste and smell), apoptosis (the process whereby the body can destroy unwanted cells), and the production of sperm.

In hindsight, the categories of genes that showed up in scans make sense given the signature of selection that was scanned for. The signature is relatively easy to detect when multiple changes have occurred over evolutionary time at functional DNA sites. For instance, genes involved in biological processes such as disease resistance are always "running" just to stay in place, like Alice running alongside the Red Queen in Lewis Carroll's *Through the Looking Glass*.[1] That is, our immune defenses are always evolving new DNA changes at functional sites. This is because, in a kind of biological arms race, the bacteria, viruses, and parasites that cause human diseases are always evolving changes in their proteins to "try" to beat our biological defenses. The end result is an easily identifiable signature of adaptation, because there have been many alterations at functional DNA sites.

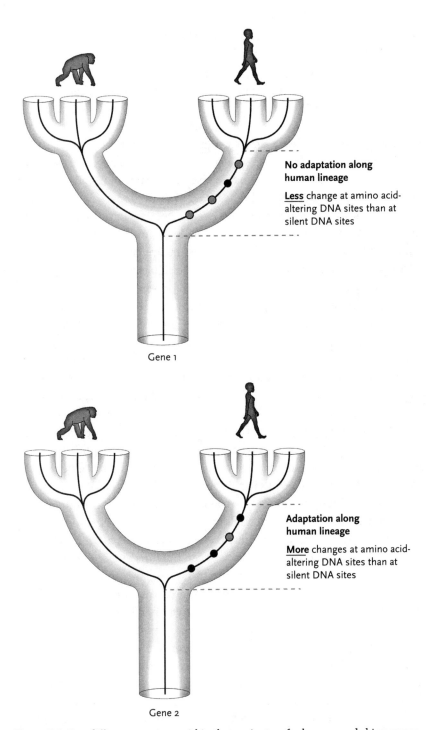

No adaptation along human lineage

Less change at amino acid-altering DNA sites than at silent DNA sites

Gene 1

Adaptation along human lineage

More changes at amino acid-altering DNA sites than at silent DNA sites

Gene 2

Figure 5.1: Two different gene trees within the species tree for humans and chimpanzees. Gene 1 shows no genetic evidence for positive selection because the rate at which functional substitutions (black dots) occurred along the human lineage has been less than at silent mutations (gray dots). Gene 2 shows evidence for positive selection because the rate of functional substitutions has been greater than the rate of silent substitutions. Note that each of the main branches that lead to humans and chimpanzees ultimately branches into smaller branches. This represents different geographic populations of each species today.

Genes in other categories like chemosensation and olfaction are also in direct interaction with changing environmental agents such as dietary compounds and smells.

The signal of positive selection in humans in a category of genes related to sperm formation (spermatogenesis) is somewhat more puzzling. In animals in which females mate with many males in the same season, females are inseminated by multiple males, and sperm from different males compete to increase their chances of fertilizing a female's egg. This is known as sperm competition. Thus a male's sperm will gain in the evolutionary sense by adapting quickly and repeatedly to out-compete another male's sperm. It is this repeated and persistent evolutionary change at functional DNA sites that is easily detectable by scanning methods and is a signature prevalent in fertilization genes in many mammals. However, it is curious that humans would show a strong signature, since the common view is that female promiscuity in human ancestors has been relatively reduced, at least when compared to our nearest cousin, the chimpanzee. Chimpanzee females are well known to be promiscuous, and can indeed mate with many or all of the males in their social group recurrently in the days when they are in maximal sexual receptivity. To investigate this finding, a recent study[2] by Alex Wong at the University of Ottawa identified several hundred genes to be expressed in the testis, and found that while in chimpanzees these genes show significantly strong signatures of positive selection, in humans they do not (in apparent contrast to previous results). So at present it is unclear to what degree humans have experienced accelerated adaptive evolution in sperm-related genes during the past. Continued work in this area should aim to clarify the specific functions of different genes—for example, whether they are related to sperm formation, sperm morphology, motility, or composition of seminal fluid—and try to identify how these sets of genes are related to differences in reproductive behavior between humans and chimpanzees. Of course, it will be useful to incorporate other primates too, since we know that the mating systems of different primates are quite variable. Among great apes, gorillas live in one-male groups where all females in the group normally exclusively mate with the dominant male and therefore expectations are that gorillas will show less evidence of sperm competition in their genomes. Indeed, the size of the testes—thought to be correlated with level of sperm competition—is very small (0.01%) in gorillas relative to body weight whereas in chimpanzees the testes are a whopping twenty-eight times larger (0.28%) and human testes fall much closer to gorilla testes' size at 0.06% of body weight.[3] In the future, detailed genomic studies should help us discover significant findings about our sexual lives in the past, and will potentially provide insights into the mating styles of our earliest ancestors.

LIMITATIONS TO DETECTING ADAPTATIONS

As mentioned earlier, when we scan across our genome in search of regions that have been shaped by positive selection, we commonly use methods that are designed to detect genomic regions where rapid and repeated changes of amino acids have taken place over evolutionary time. However, the evolution of our major human adaptations, unlike pathogen defense, probably did not evolve through rapid and repeated amino acid changes in proteins. Let's think about some of these adaptations. The past environments in which we lived did not continuously and rapidly change in such a way that required rapid and continuous readjustment in locomotor style, brain size, hand dexterity, language, and body hair. In other words, the environmental changes that drove these major human adaptations were more gradual; there were no rapid arms races as there are when our bodies need to be continuously fighting against rapidly evolving pathogens. In terms of size, *Homo erectus*'s brain remained fairly unchanged from 1.8 million years ago to 200,000 years ago.[4] This evolutionary stasis is also seen in toolmaking, where there is little change in the types of stone tools and the techniques used to produce them over long periods of time for the early members of the *Homo* lineage, *H. habilis* and *H. erectus*. The innovations we see in fossils for efficient bipedal walking occurred gradually over millions of years.

On the other hand, it is possible that changes at single amino acid-altering sites in genes could have contributed to major human adaptations. One example, mentioned earlier, would be the single amino acid change in the pigment gene *MC1R* that changes coat color in Florida beach mice. However, our large and complex adaptations are most likely influenced by numerous genes, not just single genes. And, while there may have been only a small number of amino acid-altering changes within any single gene, these changes could have been spread out over hundreds of different genes.

But the methods that most genome scans use cannot detect when a single change or a few changes to genes have caused or contributed important adaptive changes in our evolution.[5] The reason is simply that there has just not been enough amino acid change at these genes to allow the adaptation to be detectable. As we noted earlier, humans and chimpanzees diverged so recently in time (less than 5.4 million years ago) that there hasn't been sufficient time for many DNA differences to accumulate between them. In fact, about 20% of human and chimpanzee proteins are identical, and the other 80% differ by only one or two amino acids.[6] On the other hand, with so few molecular changes in proteins (and such large outward differences between chimpanzees and humans), it becomes a tantalizing possibility to consider that these few changes have had very large effects. This has led some researchers to try other approaches, especially experimental techniques, such as replacing genes in model organisms (such as mice or birds) with human genes where only a single or several DNA changes are present, to try to reveal the possible adaptive effects of these differences. In certain cases, as we will see later in the case of the gene *FOXP2*, and the role it plays in vocalization and perhaps the origins of human language, these experimental methods can prove quite useful.

The failure of genome scans to identify large numbers of genes related to our major adaptations leads us back to the important question: To what extent are our major adaptations a result of DNA changes within the coding portion of our genome? As noted, alterations to protein-coding genes may contribute to our adaptations, even if they are difficult to detect with our current methods. On the other hand, it is possible that many important adaptations reside not in the protein coding parts of genes per se, but in the short DNA regions that regulate the expression of genes.

OF MEATHEADS AND DOLPHINS

The human brain is an astonishingly large and complex organ. More than any other feature, it defines our humanity. The human brain has enabled us to develop complex languages, intricate social interactions, highly developed emotions, technological innovations, and artistic achievements. At an average mass of 1,350 grams, the human brain is more than three times larger than a chimpanzee's brain (average 380 grams).[7] Of all mammals, humans have the highest encephalization quotient (EQ—a measure of brain size relative to body weight) hypothesized to reflect intelligence. It is approximately twice that of dolphins who otherwise have the highest EQ.

But why are human brains disproportionately large and how did they get this big? The fossil record has given us more (albeit incomplete) information on when brain size increased, but it can't really answer the why and how questions. And we've wanted to know the why and how for a long time, because we think that may help to explain some of the interesting ways in which humans and chimpanzees differ in behavior and cognition. There is also a burning question we are hoping genome studies can answer, lingering ever since Darwin, about whether our cognitive differences from our nearest relatives are only a matter of degree or marked by a more dramatic shift in kind.

With the increasing availability of full genomes, neurobiologists can begin to link specific features of the human brain to specific genes, a field we can call neurogenomics. It was eye-opening when the first scans of our genome in the mid-2000s revealed few if any brain or nervous system genes showing signatures of adaptive evolution. As researchers probed further, restricting their scans just to brain genes and looking only at how the genes code for amino acids in proteins, they found that these genes were some of most unchanging elements in the genome. Thus, in large measure the proteins coded for by brain genes generally do not appear to be much different from the equivalent genes in chimpanzees. It was ironic when at least one study found that the set of human brain genes showed fewer functional changes than those in chimpanzees.[8] The study used a novel technique where the rate of evolution of the set of brain genes was compared with the rate of change at many other genes in the genome not associated with the brain. Brain genes evolved about 25% slower than most other genes in the human genome, probably because of their fundamental importance. However, while the protein structure encoded by genes may not have changed much, brain genes do appear to be expressed in increased *quantities* in humans compared to chimpanzees. As noted above, this points to gene regulation as a very important way that human brains have evolved and an area which needs much more investigation.

Some researchers have taken a different approach, scrutinizing specific genes for adaptive evolution because DNA variants in these genes are known to cause diseases or disorders in humans today. This type of study is known as a candidate gene study because the focus is on either a single gene or several genes suspected of being associated with a particular function, anatomical feature, disorder, or disease. Until now, studies have focused on less than 1% of genes known to be expressed in the brain, so there is ample opportunity for the discovery and analysis of brain-related genes.

Four such candidate genes are *ASPM*, *MCPH1*, *CENPJ*, and *CDK5RAP2*. Defects in these genes are known to be associated with the disorder microcephaly, in which individuals have a small head size and a severely underdeveloped cortex of the brain. This led researchers to surmise that these genes might be responsible for the expansion of the brain and increased cognition in human evolution. The functional link between the microcephaly genes and brains has been strengthened by the results of two different studies—one studying the genes in a Norwegian population and the other in a Chinese population—that found that DNA variants in the genes are associated with differences among people in either cranial volume or the amount of surface area of the brain's cortex.[9,10] In evolutionary studies these genes were found to have adaptive signatures. Over the course of human evolution some of the microcephaly genes experienced an increased rate of change at functional DNA sites. And accelerated change also appears to have happened earlier in our evolutionary past on the stem lineage that led to the entire ape and human group. Moreover, adaptive change in these genes does not seem to have been restricted to the lineages leading to apes and humans, but is also found more widely across the so-called higher primate group, which also includes South American and African monkeys.[11] One thing that seems to becoming clear from such broader studies is that the evolution of the human brain was primed by earlier adaptive evolution in our deeper primate ancestors, lending support to Darwin's assertion that the human brain (and mind) shows continuity with other primates, especially the apes. Strikingly, some of the microcephaly genes that appear to underlie increased brain size in humans also show signs of adaptive evolution on other branches of the primate family tree where certain species have notably larger brains.[11] This underscores another important point: that there may very well be limited evolutionary means among different species to arrive at big brains.

Big brains need an enormous amount of energy to grow and support their diverse and complex functions. Even at birth, the human brain uses about 60% of the body's entire fuel intake to support the energy needs of the brain.[12] The rate of brain growth from infant to adult is faster in humans than in chimpanzees. At birth, the human brain is more than two times larger than the infant chimpanzee's brain. By the time humans become adults, the human brain is almost three times the size of the chimpanzee brain. As we grow and learn, our brains also build connections between neurons at synapses, the locations where neurons communicate with one another. This requires the constant growth and removal of neuronal branch-like extensions known as dendrites and axons. Neurons are the most energy-hungry cells in our bodies, devouring about twenty times as much energy compared to the same amount of muscle tissue. This all adds up to a lot of energy being used by our big brains. So how do our brains get enough fuel?

In part, we have evolved efficient ways of making energy. At Wayne State University, the late Morris Goodman and his colleagues discovered that a relatively large fraction of genes involved in aerobic cellular respiration, the process whereby the body makes energy to fuel its cells, have undergone repeated bouts of adaptation in human evolutionary history.[13] These genes show repeated functional changes over evolutionary time in their protein-coding sequences, and so our traditional methods can fairly easily detect them. Some of these genes indicate that adaptation occurred uniquely along the lineage leading to humans. However, other genes evolved adaptively along the lineage that led generally to the ape and human group as a whole. This is not so surprising since apes are certainly not lightweights when it comes to brain size. Ape brains are more than four times larger than monkey brains and ten times larger than the brains of primitive primates such as lemurs. Therefore, the general hypothesis that has emerged is that big brains, whenever they occur in evolution, depend on adaptations that increase available energy for the brain. Indeed, there has been a continual increase in energy efficiency for brains with the evolution of apes and then with humans.

Another way that humans can get enough energy for their brains is to rob Peter to pay Paul, or in other words, steal energy that would otherwise go to other parts of the body. The "expensive tissue hypothesis" advanced in 1995 by anthropologists Leslie Aiello at University College London and Peter Wheeler of Liverpool John Moores University proposes that during human evolution we reduced the size and elaboration of our digestive systems so as to provide more energy to our brains. It is obvious our guts are reduced in size compared to other primates and would require less energy to grow and maintain.[14] Following the rationale of the hypothesis, it is striking that genetic evidence has been discovered that shows that the genes coding for the transporter proteins that carry glucose to our brain are up-regulated (meaning more of these transporters are made), while genes coding for transporters that carry glucose to our skeletal muscles are down-regulated (meaning less of these are made).[15] These changes are all carried out through gene regulatory changes. In fact, we are finding that energy genes in the human brain are extensively up-regulated compared to energy genes in the chimpanzee brain.

As we develop from fetus to adult, the human brain may call on different sets of genes, each set adapted for different functional roles. For example, adult human brains show increased expression of energy-related genes. The brains of human fetuses by contrast show increased expression of genes involved in neural development and neuronal connections.[16] In very early development we appear to be mostly building the special structures and elaborate connections of our brains, while as adults we appear to be mostly trying to sequester enough energy to efficiently run these expensive machines.

If the hypothesis is true that the evolution of energy genes goes hand in hand with the evolution of big brains, then other animals that also have big brains also should show adaptations in energy genes. African elephants and bottlenose dolphins both have relatively very large brains compared to many other mammals, and also have some other human-like traits such as high intelligence, strong social bonds between individuals, lengthened development, and long lives (Figure 5.2). At Wayne State University, Derek Wildman and colleagues discovered in separate studies in 2009 and 2012 that sets of energy-producing genes in the African elephant and bottlenose dolphin genomes also showed evidence of adaptive evolution.[17,18] For the elephant, this finding is made more striking when the elephant genome is compared with the genome of its cousin, the lesser hedgehog tenrec from the island of Madagascar—which looks really nothing like an elephant at all. Tenrecs have comparatively very small brains, faster development, and much shorter lifespans (Figure 5.2). As one might surmise, tenrecs do not show evidence of adaptive evolution in energy-related genes. In the future, it will be important to continue comparing genomes between different organisms to gauge the genetic underpinnings of how different species adapted to unique ways of life. It may also help us to discover ways in which some of our genes may have evolved similarly to genes in unrelated organisms, with which we share some similar adaptive features.

Human fossils indicate that there was a large increase in brain size—about 1.8 million years ago when the hominin species Homo erectus first emerged. Before this, our human ancestors had brains that were not too much larger than the brains of chimpanzees and gorillas. Various anthropologists have hypothesized that a shift toward eating meat and fat provided the fuel that enabled the evolution of larger brains by meeting the energy requirements of the increased brain tissue. In other words, eating meat lifted an energy constraint that would have otherwise put the brakes on brain size. Harvard University primatologist Richard Wrangham, in his book Catching Fire, suggests that cooking meat and other foods generally increases their energy content and makes them easier to chew and digest.[19] Some of the first archeological evidence pointing to the use of fire is associated with Homo erectus (although there are differing opinions on this), and this hominin's jaw and tooth sizes suggest that it applied reduced chewing forces. Homo erectus' more linear body form indicated by its skeleton also suggests a digestive tract appreciably reduced in size from that of its ancestors. As mentioned in Chapter 4, a gene (MYH16) linked to the development of chewing muscles developed a mutation about the time of early H. erectus, causing these muscles to become weaker.[20] This weakening may have allowed expansion of the brain at this time. Such hypotheses may be true, but the necessary work needs to be done to verify the function of this gene, still poorly known.

Figure 5.2: The photos of the African Elephant and Bottlenose Dolphin are both pictures of mother and child. These mammals possess relatively large brains and have strong family bonds and complex social behavior similar to humans. The Lesser Hedgehog Tenrec is much smaller-brained and has less complex social behavior. Comparing our genome with genomes of other species should help us find the genetic bases of traits for which we are similar to other species and traits for which we are different from other species. (Photo credits from top: Mustafa Mohammed, Steve Shippee, and Jeff Whitlock).

One of the products of digested meat is the chemical creatine, which the body can then turn into a high-energy compound by putting it through a special metabolic pathway whereby several genes work together. One study found that two crucial genes within this pathway show substantial up-regulation in the human brain compared to the chimpanzee brain.[21] Thus while meat consumption may have enabled brains to increase in size, it was also necessary for our chemical machinery to evolve in order to maximize the amount of energy we could get from meat. In the future, candidate gene studies should target other proteins, especially enzymes, involved in meat digestion and metabolism, to search for signs of adaptive evolution in human ancestry.

Greater consumption of meat by humans compared to great apes presents something of a conundrum. Studies of animals in laboratories have consistently indicated that greater fat, cholesterol, and caloric intake—everything that early humans would get more of when meat became a focus of their diet—is associated with accelerated pathogenesis and shortened lifespan. In fact, when chimpanzees and gorillas are fed diets high in meat and fats, they appear to be particularly susceptible to hypercholesterolemia and vascular diseases compared with humans. In nature, gorillas eat vegetation and chimpanzees are omnivorous, consuming meat very rarely. So how did human lifespan increase beyond the lifespan of our ape cousins, given our species' taste for meat? How did we avoid the diseases brought on by high fat and cholesterol in our diets?

One intriguing hypothesis has been raised by the neuroscientist Robert Sapolsky at Stanford University along with Caleb Finch and Craig Stanford at the University of Southern California.[22,23] They suggest that as humans began to rely heavily on meat during human evolution, they evolved so-called "meat-adaptive" genes. These genes allowed humans to live longer despite a high intake of cholesterol and fats. One candidate gene is the Apolipoprotein E (*APOE*) gene. While one version of the gene—known as the E4 allele—is common in apes and monkeys, the E3 version of the gene evolved in the genus *Homo* and is the most common form of the gene in humans today. There is some evidence that the E3 allele protects against vascular disease and cognitive diseases like Alzheimer's disease. Other genes are thought to mediate fat metabolism and thereby lessen our susceptibility to chronic diseases of aging. What's now referred to as the Finch-Sapolsky-Stanford Hypothesis has suggested far-reaching consequences of the evolution of the new *APOE* gene. According to their hypothesis, it may have allowed humans to lead longer reproductive phases of their lives; to increase in overall lifespan by reducing cognitive and vascular disease, thereby permitting more extended care of young by parents and grandparents; and also to produce larger brains through the increased consumption of meat in the diet. This hypothesis needs much further testing to verify the exact function of the new *APOE* gene in humans, and to determine if and when "meat-adaptive" genes evolved in the human genome. But if "meat-adaptive" genes are found, and we can determine when they first started to evolve, we should be able to finally obtain answers to some of our persistent questions: when in our evolution did we first start eating meat in a big way, does it correlate with the first archeological evidence for hunting large game, and how did meat-eating influence our bodies, brains, life spans and social lives?

GENE REGULATION AND THE HUMAN BRAIN

Before the genomic age, in 1975, the molecular biologists Mary-Claire King at the University of Washington and the late Allan Wilson raised the important hypothesis that many important differences between humans and chimpanzees may be due to evolutionary changes in regulatory regions of the genome.[24] They were led to this conjecture because comparisons among proteins of the two species were revealing they were almost identical. It should be noted that at the time they had little idea where such regulatory regions were located in the genome or how they worked. What they did surmise is that these regulatory DNA regions could influence genes by either increasing the expression of the genes to produce more protein product, or by decreasing the expression of the genes so as to produce less protein product. Regulatory regions might "turn up" a gene in one tissue like the brain, but "turn down" this gene in another tissue like the liver. This hypothesis is even more relevant today because we now have a better understanding of how regulatory regions can operate to control genes and we know where some of these regions are located within the genome. And we now also have massively more data than in 1975 with which to examine the validity of this hypothesis.

The approximately 1.5% of our genome that contains genes leaves the vast majority of our genome as a potential source of functional material. However, analyses thus far have found that much of the remaining genome is composed of "junk" DNA, a term coined by the late geneticist Sasumu Ohno.[25] To qualify the meaning of this description, "junk" DNA is understood to be DNA that does not encode information important for the survival and reproduction of the organism. Indeed, recent genomic comparisons have discovered that only up to roughly 5% to 10% of the "non-gene" DNA may in fact possesses important functional roles, most likely in regulating the expression of genes.[26] Figuring out how all these regulatory elements interact with genes and how they contribute to human adaptations provides fodder for much future research.

So what have we discovered about regulatory regions so far? We know that gene regulatory regions are generally very short, consisting of less than fifty bases of DNA sequence. We also know that regulatory regions are often located near the genes they regulate, either just before the gene or within "spacer" regions known as intron regions that occur within the gene. However, the ENCODE project (Encyclopedia of DNA Elements) is discovering that many regulatory regions are located far from the genes they regulate, so identifying them becomes much more difficult. For example, the promoter region of the lactase gene—which functions to regulate the gene and influence whether or not its product, the lactase enzyme, is produced—is about one hundred bases long and is located almost 14,000 DNA bases away from the actual lactase gene, falling instead within a gene having an altogether different function, a function in cell division.[27]

How do we know if regulatory changes contributed to human adaptations? One way is through gene expression studies. Such studies measure differences between closely related species in how much of a gene's product is expressed within a specific biological tissue. These studies usually target an anatomical region (e.g., the liver, the kidneys or the brain) and measure amounts of messenger RNA (mRNA) in that tissue (since a gene is transcribed into mRNA in the first of two steps towards producing an actual protein). One field of inquiry where gene expression studies have offered some intriguing insights is the uniqueness of the human brain. Studies survey sets of genes and measure whether or not there are differences in amounts of mRNA for these genes between the brains of humans and chimpanzees. Indeed, compared to chimpanzees, gene expression in humans has been found to be generally increased in the neocortex, the outer layer of the brain associated with higher brain functions like sensory perception, spatial senses and reasoning, initiating motor commands, conscious thought, and language.[28] A recent study noted a particular increase in expression of genes in the frontal part of the neocortex,[29] a region enlarged in recent human evolution and involved more specifically in social cognition, forethought, and abstract representation. The task then becomes to identify exactly what these genes do to our brain. In the frontal neocortex, there is evidence that the genes increase connectivity among different nerve cells by increasing numbers of neuronal axons, spines, and dendrites; however, other important functional changes remain to be found. Infant and juvenile brains of chimpanzees and humans have also been compared to gain insight into brain differences in the phase of life when learning is most intense. Preliminary findings show that young human brains show patterns of gene expression that are relatively immature compared to chimpanzees at the same age.[30] This lends genetic support to the idea that, during the course of human evolution, humans have become progressively slower to mature compared to chimpanzees and other primates—a process called neoteny ("to stretch the features of a young individual")—with the consequence that our brains remain malleable longer, allowing us more time to engage in learning.[31]

A similar protraction of development in humans has been found in the myelination of neurons of the cortex. Myelination is the process whereby the long axon branches of neurons accumulate the fatty layer that speeds up nerve impulse propagation. Neurons in the human cortex are found to take significantly longer to develop these "fast tracks" than chimpanzee neurons.[32] In fact humans are born with significantly fewer myelinated neurons than chimpanzees and whereas chimpanzees reach full myelination at sexual maturity, humans take their time, only reaching full myelination well after adolescence. Thus, humans are still building their neural circuitry as they are growing up within a complex cultural environment where intense learning is taking place. This helps to produce our more complex cognitive functions. In the near future, we anticipate that we will learn much more about how gene regulation has contributed to the evolution of our brain and other adaptations. We are only at the beginning of understanding how changes in regulation of genes may have produced chimpanzee-human differences such as those we described in the brain. It will be necessary to go several steps further to identify the actual genes responsible for the differences, and exactly how many and which DNA changes have caused the differences in how these genes are regulated.

THE SKIN WE'RE IN

Ever since Desmond Morris argued in *The Naked Ape* in 1967[33] for an explicit sexual explanation for nakedness, researchers have been studying the significance of our virtual denudation. Among all the primates, only humans do not have a dense coat of fur. When did we first lose significant amounts of our body fur? What were the possible adaptive benefits of losing our fur? Do we know what genes underlie our nakedness?

If you were able to tease back the fur of chimpanzees, or any other non-human primate, you would see light skin beneath their fur. Humans, in contrast, have an enormous variety of skin colors, with Africans (considered to be the oldest human population) having mostly dark skin colors (even though considerable variation in dark color exists amongst different African peoples). Although multiple genes are now known to underlie skin color, one key gene under investigation is MC1R (melanocortin 1 receptor protein), which can act as a switch between whether a dark or light protein pigment is produced in skin. The African MC1R is thought to have evolved to protect against ultraviolet light (UV) skin damage from the sun. Among African people there are no amino acid differences in MC1R, an unusual case because we would expect there to be more genetic variation among Africans due to their long evolutionary history. The lack of variation is believed to be due to the ever-acting force of negative selection removing any amino acid mutations that might have arisen over evolutionary time that would lead to lighter skin color and harmful effects from intense UV exposure.[34] Europeans and Asians present a different case, because we find plentiful amino acid variation and much of this variation causes the M1CR gene to lose its function and leads to lighter skin colors. One hypothesis explains that the need to maintain a dark protective skin color was not as important in people that had expanded from Africa to regions in Europe and Asia, where UV light is less intense.[35]

Interestingly, there are also many amino acid differences in the MC1R gene below the light skin of chimpanzees, as in Europeans and Asians. We believe that chimpanzees have a low risk of skin damage from high UV light because chimpanzees have dense fur and live primarily in for-ested environments. But, somewhere early in our evolution we moved into more open environments, with little to no tree cover, and were then exposed to the intense equatorial sun. So what was the color of the skin of our earliest human ancestors? Did our early human ancestors still retain their fur? It seems quite likely that our earliest ancestors inherited a luxurious coat of hair from their shared ancestor with chimpanzees and therefore would have been shielded from the UV light. It makes sense then that they likely also maintained the light skin color of their primate cousins.

What can all this tell us about when we "naked apes" started to lose our dense body hair? At some point in human evolution, MC1R appears to have evolved several amino acid changes that shifted skin color from light skin toward the darker end of the spectrum. These changes must have been adaptive because they spread to all Africans, and after they were in place, negative selection prevented any further mutational change in the gene because it was very important to maintain a dark skin color. In a clever insight, Alan Rogers at University of Utah[36] determined that if it were pos-sible to estimate the time when the African MC1R gene first became con-strained by negative selection, this might also suggest when we started to lose our hair. After all, evolving and maintaining a dark skin color would only be necessary if, so to speak, the fur had already flown. But how to estimate this time?

Although the *MC1R* gene shows no amino acid differences between African individuals, Rogers knew that the gene would have differences between individuals at the "silent" DNA sites we discussed earlier. Since mutations at these sites are believed to have occurred regularly through evolutionary time, it is possible to use the amount of differences observed among African individuals at "silent" DNA sites to estimate the time when *MC1R* became constrained. Using this method, Rogers estimated that humans have been dark-skinned since at least 1.2 million years ago. If we were already dark-skinned a million years ago, this means that there must have been a selection pressure favoring persons with darker skin because it protected them from the harmful effects of UV light. But such a selection pressure would arise only if we had already lost our body fur. So in fact, 1.2 million years becomes a minimum date for when we lost our body fur, and it is likely it occurred much earlier in our evolution.

We can also learn about hairlessness by analyzing genes of an organism very well acquainted with human body hair—body lice! Humans, unlike chimpanzees and gorillas, can be infected by two species of body lice. Besides the species that can infect the hair of the head and body, there are lice that specifically infect the more isolated pubic region. The species of head/body lice that infects humans is most closely related to chimpanzee lice. The chimpanzee louse and human head/body louse made their evolutionary split near the same time that humans and chimpanzees went their separate ways (five to six million years ago). Therefore, it seems that our intimate parasites separated from each other at the same time their hosts separated.

In a fascinating discovery in 2007, David Reed at the Florida Museum of Natural History[37] found that the louse that colonizes the human pubic region is not most closely related to chimpanzee lice, but instead to gorilla lice. Strangely, however, the time of divergence between human pubic lice and gorilla lice is estimated to have occurred three to four million years ago, a time that is millions of years more recent than the time that gorillas and humans made their evolutionary split (around eight to ten million years ago). What could account for human pubic lice being so closely related to gorilla lice?

The best explanation seems to be that gorilla lice were "passed" to humans at a much later time compared to when these two species diverged in evolution, perhaps through some kind of contact between the two species. In humans, pubic lice are spread between individuals primarily through sexual contact. While it is titillating to think that humans and gorillas might have engaged in cross-species sexual activity, it is more likely that humans came into contact with gorilla lice through some other means, like hunting or scavenging gorillas. Also, since gorillas build leaf nests to sleep in, early human ancestors may have investigated or otherwise came into contact with gorilla nesting sites thereby catching gorilla lice.

While the three- to four-million-year date for the separation between gorilla lice and human pubic lice can't tell us anything for certain about human hairlessness, in order for the gorilla lice to invade and then thrive in the human pubic region, the human pubic region must have been relatively accessible at this early time in human evolution. By accessible we mean that there must have been no other competing lice species in the pubic region, such as the human head/body lice species, that could have prevented the invading gorilla lice from taking hold. This seems to indicate that by three to four million years ago humans had already lost a continuous coat of fur connecting the pubic region with other regions of the body. In other words, the pubic hair was already isolated from our other body hair, and it's likely that humans had already become considerably less hairy than their ape relatives.

Although we have obtained some clues as to when we lost our fur and how human skin color likely went from light to dark as a consequence, we still lack information about how many changes in our genome were needed for the gradual loss of fur over our bodies. Another curious feature of humans, for which we do not yet know the genetic underpinnings, is the ability to sweat over the entire skin's surface. In most primates and other mammals, "sweating" is only through the mouth (i.e., panting). Although we can speculate that sweating evolved conjointly with loss of fur—loss of fur enabling early human ancestors to be active in sunny, hot, and open environments like the African savanna and full body sweating providing efficient cooling and enabling sustained activity patterns—it is impossible for us to know. Indeed, the selective pressure on early hominins to thermoregulate efficiently is a major theme in explanations for the origins of bipedalism and evolution of larger brains, since brains are very heat sensitive.[38-39] Research by Daniel Lieberman supports the idea that it was endurance running by *H. erectus* during hunting (ca. 1.9 million years ago) that selected for our fur-loss and enhanced sweating abilities, since endurance running is more intensely heat-producing than simply walking.[40] Once the genetic systems underlying both fur-loss and increased number of sweat glands are pinpointed we may gain a better idea about when these two features evolved, their links to endurance running and the evolution of big brains.

THE LANGUAGE OF MICE, MEN, AND BIRDS

One remarkable aspect of human cognition is language, the ability to organize meaning into sounds and to share our thoughts with others through sometimes subtly varying vocal sounds. Classic language studies in apes have shown that while they are able to learn the meaning of certain language sounds, apes are unable to organize words into strings of meaning and lack the ability to produce the varied sounds of human speech. Somewhere along the evolutionary lineage leading to humans we developed these unique and extraordinary abilities, and this would have vastly increased our capacity for learning and accelerated both our cultural evolution as well as our brain's evolution. Some biological anthropologists believe that language evolved only recently when our species, in its anatomically modern form, evolved and moved into Europe about 40,000 years ago, and that language was not part of the behavioral repertoire of Neandertals, our archaic cousins who lived in Eurasia between about 300,000 and 30,000 years ago, nor of our earlier ancestors.[41] Others have hypothesized, based on their interpretations of archeological evidence, that both Neandertals and possibly our earlier ancestors in Africa had already evolved human-like language abilities.[42]

Part of the idea that language is a uniquely modern human behavior and that it was acquired only recently is the notion that a discrete genetic revolution took place in the evolution of modern humans that led to the neural machinery needed for language. However, at present this is a rather vague idea, especially because we have little awareness of how many genes underlie language or where in the genome these genes might be located. It is also not straightforward to tell merely from archeological remains (e.g., stone tool complexity, carved objects and burials) whether or not a particular human ancestor had language abilities, and if they did, to what degree. It is also difficult to imagine that language did not develop in a gradual way and that less complex forms of language didn't exist in a sort of evolutionary progression, at least within the last several hundred thousands of years.

Recent neuroimaging studies of subjects' brains, whereby specific areas of the human brain are imaged while performing specifically designed tasks, indicate at least two general language-related neural networks in the cerebral cortex.[43] One network is related to mapping meaning onto sounds and the other is involved in linking the meaning of sounds to their articulation in speech. These studies have revealed that these networks interconnect various regions of the cortex and not just the most well-known language-specific regions, Wernicke's and Broca's areas of the cortex. Such a complex system undoubtedly involved genetic changes within numerous genes. Though we are only at the very beginning of understanding the genetic basis of our language abilities, at least one gene, known as *FOXP2*, has so far been discovered that plays a substantial role in human speaking abilities, and has likely played an important role in the evolution of human language.

Mutations in the gene *FOXP2* in humans are known to cause severe impairments in the production of speech as well as in language comprehension, probably due to its link to defects in parts of the brain that coordinate the muscles of the larynx (voice box) and mouth as well as in areas of the brain that help generate symbolic meaning. In species like birds and mice, the gene is involved with generating vocalizations. In zebra finch birds, damage to this gene impairs their ability to imitate and learn songs. But the signature of selection at *FOXP2* in humans is not the type that would easily be detected in genomic scans. The gene has undergone only two amino acid changes along the human lineage. Such a pattern of change would remain undetected by the usual methods of detecting selection that require greater numbers of amino acid changes. One aspect that makes the gene stand out is that the *FOXP2* protein has remained almost unchanged over the entire evolution of mammals (about sixty-five million years or more). A change of two amino acids along the human lineage, less than 8% of the timeframe of mammalian evolution, is therefore a most tantalizing discovery.

The scientific pursuit of *FOXP2* has entailed some very fine experimental work that further implicates this gene in the story of human language. In one study, by Wolfgang Enard and colleagues at the Max Planck Institute in Leipzig, Germany, the human *FOXP2* gene with its two human-specific amino acid changes was introduced into mice embryos.[44] Young mice pups with the human gene, known as "humanized mice," exhibit several very notable changes in their ultrasonic vocalizations. They didn't stand up and say "more seeds please!" but were found to exhibit new and different vocalizations than the usual calls they make when isolated from their mothers. Humanized mice also showed decreased levels of the neurotransmitter dopamine and increased synaptic flexibility of the neurons of the basal ganglia, structures buried deep in the cerebrum that in humans are involved in articulation of speech and in interpretation of the meanings of words. In young birds, the basal ganglia and dopamine are crucial for learning songs, and in humans it seems *FOXP2* may play an important role in language acquisition and in learning to control the complex coordinated face and mouth movements associated with speech.[45] In the "humanized mice," additional modifications were found in the cerebral cortex, especially in its connections to the basal ganglia, indicating that the *FOXP2* gene causes extensive and significant changes in interacting parts of the brain that are involved in language. Thus experimental evidence strongly implicates *FOXP2* as one gene that contributes to human language—that is, short of producing a Stuart Little!

Experimental research of the kind performed on *FOXP2* provides a model for the deeper investigation of other gene regions believed to underlie human adaptations, but for which we have little idea of the gene's actual function. Despite these initial findings on *FOXP2*, there is no doubt there is much more to learn about this gene, as well as other genes involved in human language. For example, we know that *FOXP2* is expressed in regions of the brain that are presumed not to function in language, in regions of the body outside the brain such as the lungs and other tissues. (However, the lungs in humans play an important role in regulating the air pressure changes over the vocal cords necessary for speech.) If *FOXP2* is indeed a gene that has contributed uniquely to the evolution of human speech, it is a fascinating point that this gene is actually a regulatory gene, meaning its protein influences the expression of an extensive network of genes in the brain. The evolutionary neuroscientist Todd M. Preuss at Emory University has suggested that *FOXP2* "might play a very specific role, for example, by orchestrating a whole set of genes that switch brain development from an ancestral program to a human program that causes cells and connections to differentiate into systems that sustain speech or language. It might even regulate the development of other parts of anatomy, such as the lungs and larynx, involved in speech production" (p. 10703).[46] Thus, the *FOXP2* gene might be an example supporting King and Wilson's proposal that many human adaptations rest upon changes in regions of the genome that regulate the expression of other genes in the body.

Speaking requires a listener, so adaptations in hearing likely evolved with the ability to speak. Several genes related to hearing have undergone functional changes in their proteins, which have signs of having been adaptively driven.[47] One gene with a strong signature of adaptive evolution is *TECTA*, a gene expressed in the tectorial membrane of the cochlea, the fluid-filled organ where vibrations trigger "hair" cells (special neuronal endings) associated with different sound frequencies. Although the exact functional significance of the protein changes in *TECTA* are unknown, people with abnormal *TECTA* genes are known to suffer from high-frequency hearing impairment.

Several other genes implicated in hearing function showed up in some early genomic scans because they had functional changes in their proteins that indicated adaptation.[48] Most of these genes are expressed in the ear or are more finely associated with "hair" cells that trigger nerve impulses at different frequencies. Most of the genes are also associated with some form of hearing loss in animals or in humans. Unfortunately, almost all of the genes are associated with other conditions as well. For example, the *EYA1* gene is also associated with eye loss in fruitflies (in fact, *EYA* stands for "eyes absent"). So it will take detailed experimental work, equivalent to what has been done with *FOXP2*, to nail down the functional effects of the changes in these proteins. Nevertheless, it would appear that fine-tuning of hearing acuity was necessary for individuals to properly distinguish and then understand the meaning of various tones of spoken language. In initial brain-imaging studies, the microcephaly genes discussed above (particularly *ASPM*) have been found to be associated with the ability to distinguish linguistic tones.[49]

The genetic basis of hearing abilities in relation to the tones of music is another area that could be closely investigated in the future. The first known musical instruments were small flutes, from the Geißenklösterle Cave on the banks of the Danube River in southern Germany, made of swan bones and mammoth ivory tusks and dating back to approximately 42,000 years.[50] Comparing genetic differences between people who have perfect pitch (able to sing any note without a reference note) with persons that can hold a tune, and then with those folks who are basically tone-deaf, may enable us to find more genes that are important for detecting the nuances of vocalizations. (Such genes may also prove important in self-listening to ensure you're producing the tones of speech accurately enough to be understood by others.)

GENOMIC EXCAVATIONS OF BIPEDALISM

Most of our unique human adaptations like bipedalism, jaws that no longer jut forward, increased hand and finger dexterity, reduced size of the anterior teeth—especially the canines—and our brains involve complex sets of functionally interrelated morphological features. Moreover, if we break down these adaptations we would likely find that their parts contain small but measurable differences from individual to individual within living populations. In this way they are similar to characteristics like height, body mass, and skull shape, all of which show a spectrum of measurable variation. And as the lab results from any blood work reveal, other characteristics like blood pressure, blood lipid levels, metabolic rate, and glucose levels demonstrate a normal range of variation associated with each complex trait measured.

While genomic studies will improve our understanding of how these features are encoded in our genome, we should be leery of thinking single genes, or even several genes, influence these traits. In fact, studies of laboratory organisms like fruitflies, fish, and mice indicate that multiple and often many genes influence these complex traits. Finding all these genes and assessing the magnitude of their influence on these traits will be a major challenge for future research.

We have also learned from laboratory studies that genetic alterations underlying complex and continuously varying traits often occur within regions of the genome that regulate how genes are expressed, rather than within the protein-coding portions of genes. The adaptations in our pelvis, lower limbs, and feet that allowed us to walk bipedally and those modifications in the shape of the bones making up our hand that facilitate the careful manipulation of objects could have arisen from shifts in the timing and locations within the body where genes are expressed. We may take as an example the discovery that regulatory DNA changes in stickleback fish influence whether the bones and spines supporting their pelvic fins develop their full size or are reduced or absent.[51] We have evolved much shorter and wider pelvic girdles compared to chimpanzees, whose pelvises are long and narrow. Could our evolved modifications to the pelvis be underlain by similar DNA regulatory changes to those seen in stickleback fish? As another example, molecular studies have found that the different sizes and shapes in the beaks of Darwin's finches are governed by regulatory mechanisms that increase or decrease the protein product of specific genes.[52] Could our smaller and pushed-in jaws have evolved through mechanisms similar to those that drove the evolutionary changes to finches' beaks? If regulatory changes played a large role in the evolution of major human adaptations, then their genetic basis will prove more challenging to detect. This is because these changes do not reside within the regions of our genome we know best and for which our genome-scanning methods have been traditionally designed, namely the coding parts of genes. Luckily, experimental techniques and manipulations of genes in model organisms (e.g., birds, fish, and mice) may help us to locate human-specific regulatory changes that underlie our major adaptations.

SUCCESSES AND CHALLENGES

Despite challenges in excavating our adaptations in the genome, we have already enjoyed some success. We have discovered several genes that appear to contribute to our unique brains by influencing neuron shape or by helping the proliferation and differentiation of brain cells during development. We have also detected many genes that appear to have undergone adaptation to increase the energy supply to our brain, and we have found at least one gene that contributes to human language, though others most certainly will be found. Moreover, the success of gene expression studies has shown us that certain sets of genes are very differently expressed in human brains compared to chimpanzee brains, and that some molecular differences seem to be related to the prolongation of development of the human brain, perhaps allowing our children to develop their brains over a longer period of time in rich experiential contexts.

However, despite initial successes of expression studies, we still need to maintain a degree of skepticism in our results. This is especially so since mRNA, as mentioned above, remains a step away from the final protein product, and new research has shown that differences in levels of mRNA between species are not always accurate indicators of differences in levels of protein product present in a particular tissue, and therefore may be poor indicators of an actual phenotypic difference.[53] Thus future investigations will need to assess the accuracy of initial gene expression discoveries—as well as explore what kinds of regulatory processes take place subsequent to the first step of the gene-to-protein process that can derail the association between mRNA levels and of amount of protein product expressed—so that we can advance toward understanding human-specific adaptations.

Increasingly, we will need more experimental approaches such as those used in the FOXP2 studies to determine the effects of amino acid alterations within proteins. And, since some of the genes that underlie our human adaptations play regulatory roles, we will need to carefully map out the network of other genes in our genome that are turned on or off by these "master" genes. Indeed, for any gene discovered with a genetic signature of selection, experimental work is necessary to verify if the DNA changes in this gene really have a measurable effect on appearance and behavior. Without this kind of careful investigation, our stories of adaptation in human evolution will be no more than Kipling's "just so" stories.

To better understand the genetic basis of our species-wide adaptions, we can also look deeper into the past, to see how our adaptations have emerged from our more ancient evolutionary history among the primates. The fact that more and more genomes from primate species are becoming available makes it entirely possible to trace the evolutionary stages of our human features back in time and to realize that some of our major adaptations like our brain size were primed long before the human lineage appeared nearly five million years ago. Already we have discovered that some of the same genes influencing brain size and function in humans were under positive selection in earlier primates. We are discovering genetic support for Darwin's prescient conjecture in *The Descent of Man* of the biological and qualitative continuity in mind between humans and our primate relatives. Comparing the human genome with genomes from more distant mammalian relatives, like elephants and dolphins, indicates that sometimes even these animals have evolved some features in genetically similar ways to humans. With respect to language, it is most intriguing that the same *FOXP2*-regulated brain circuitry involved in helping young birds learn to sing their species-specific songs also appears to help young babies and children learn their specific languages. Finally, we can study biological features that vary among people in our modern populations. After all, we have much more evidence about the present than the past. If we can better understand the identity and number of genes that underlie variation among living peoples in skin color, hair texture, and head and body shape and stature, among other features, this information could hint at the genetics behind "species-wide" human adaptations.

The genomic age will help us get closer to answering age-old questions of who we are and how we came to be, and at the same time will help reveal our evolutionary connections with other life on Earth. In the next chapter, we will explore how genomic evidence can be used to study the evolutionary origin and geographic expansion of modern humans.

CHAPTER 6

The Genomic Origins of Modern Humans

As we've seen, our genome is like an evolutionary patchwork quilt. It contains many thousands of independent DNA segments shuffled between the sperm and eggs of fathers and mothers from the time of our separation from the apes until today. These independent recombining bits of our genome are a huge asset to scientists in developing accurate reconstructions of our past. They were essential in allowing us to prove that chimpanzees and bonobos are our closest living primate relatives, and they are equally important for developing an accurate understanding of the origin and evolution of modern humans—humans with fully modern features like you and me—within the last half million years or so.

Many of the questions we wish to answer about evolution concern phenomena not at the level of a single gene or individual, but at the population level. The speciation process between humans and apes, leading to our reproductive isolation from chimpanzees and bonobos, is one example of a population process that had the effect of producing many similar gene trees across thousands of genes across the genome. Most of our questions about the origins and evolution of our modern species are likewise questions at the population level, where many segments across the genome are expected to show similar patterns. Consider the following series of questions. Where was the geographic origin of our modern species? How long ago did we originate? Was the process of our origin spatially and temporally discrete or gradual and spread out across more than one ancestral population? How did populations of modern humans colonize different regions of the world? Did our rapid expansions into new territories have effects on the human genome? How, when, and where along our evolutionary journey did populations dwindle or expand in size? Answering these questions relies on the search for congruent patterns among the many gene segments that make up our genome.

To address these many questions, researchers are determining the full DNA sequences of entire genomes from people around the world. Like detectives solving a crime, the best way for researchers to reconstruct our evolutionary past is to look for bits of independent evidence that all point to the culprit or culprits and to the same timeline of events. Until recently, however, many such studies focused only on a single piece of DNA evidence, either the small circular mitochondrial genome or the Y chromosome. While studying these systems offers unique benefits, and continues to yield important insights about our past, they are severely limited in several ways. First, the sizes of our mitochondrial genome and Y chromosome are only minuscule fractions compared to the size of our nuclear genome. For instance, if a DNA base were equal to the distance of a foot, the full nuclear genome would stretch to the moon, back to earth and then back to the moon again, yet the Y chromosome would only make it about 5% of the distance to the moon. The mitochondrial DNA genome (only about 16,500 bases in length) would only reach the top of Mount Kilimanjaro in Tanzania. Second, both of these genetic systems are inherited from a single parent, each one only through the maternal or paternal lineage, and so each system can only potentially give us information about the evolutionary history of females or males. This could very well be different from the evolutionary history of our entire species.

But by far the most significant limitation of studies based solely on the Y chromosome and mitochondrial genomes is that they do not recombine genetic parts; that is, in the formation of offspring, mitochondrial genomes and the Y chromosome (by and large) do not shuffle together different segments inherited from the parents. Thus, they do not offer the opportunity for scientific testing, through the search for congruent patterns across independent segments, which would support or refute our evolutionary hypotheses. Another very significant problem is that genomes without recombining segments can yield gene trees that are easily distorted by natural selection, thereby reducing their usefulness for tracking the history of a species or population. For example, in the process known as a selective sweep, described in detail in the next chapter on population adaptation, when a beneficial DNA site rapidly spreads through a population due to positive selection, other DNA sites in the same linkage group as the beneficial DNA site also tag along for the ride. In the nuclear genome, these DNA tag-alongs are found only in the relatively short region around the beneficial variant—that is, within one out of a huge number of distinct linkage groups—but in the mitochondrial genome and Y chromosome, DNA variants over the whole genome or chromosome will be affected because they all belong to a single large linkage group. Because of these limitations, hypotheses about our evolutionary past based solely on analysis of genomic systems like mtDNA and the Y chromosome have the real potential to be misleading.

Over the course of twenty years, the accumulated evidence from mtDNA and Y chromosomal genetic variation from diverse world populations increasingly supports the hypothesis of a recent African origin of our species. Africans show the oldest mtDNA and Y lineages, and African populations show the greatest amount of genetic diversity compared to populations outside Africa. All mitochondrial variation found in humans coalesces to about 200,000 years ago[1,2,3] and to about 100,000 years ago for the Y chromosome,[4] which has led many researchers to support a relatively recent origin for our species. Yet many questions still remain—not the least of which is how hypotheses about the location and timing of modern human origins that are based on two small uniparental genetic systems fare when the vastly larger nuclear genome is studied. Luckily, the nuclear genome, with its many independent genomic segments, offers unprecedented power to test these hypotheses, and is already allowing us to dig deeper and uncover answers to many questions we have about our past.

AFRICAN ORIGINS

The most common genetic differences studied between populations are differences between individuals in the A,C,G, or T bases that make up the ladder of the DNA code, known as single nucleotide polymorphisms, abbreviated SNPs, and pronounced as "snips" (Figure 6.1). (The word *polymorphism* simply refers to differences that exist between individuals of a species.) We have learned that, when we compare two random individuals, SNPs exist at a rate of about one DNA difference for every 1,000 bases. By contrast, human and chimpanzee genomes differ in this way by a factor of ten—that is, about one base differs for every 100 bases of DNA. Of course there are other differences between the genomes of different individuals— whole duplications of genes, deletions of bases, insertions of bases, and others—but with the exception of duplications, we will be mostly focusing our discussion on SNPs.

Recent studies of the full nuclear genome have offered some of the most convincing support for our evolutionary origin in Africa. These studies

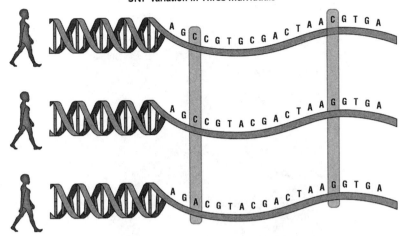

Figure 6.1: DNA sequences compared among three different people that illustrate single nucleotide polymorphisms (SNPs). The third individual has an A nucleotide at position 3 whereas the other two individuals have a C. The first individual has a C nucleotide at position 15 whereas the other two individuals have a G. Each of these sites represents an SNP.

reveal that genetic variation is highest in Africa but shows sequential reductions in genetic diversity as populations are sampled from increasing geographic distances from Africa. In other words, populations today that live farther and farther away from Africa contain ever-decreasing subsets of the genetic variation that we see in Africa. This decline in diversity is due to the so-called serial founder effect, a process we believe is the result of successive expansions by small groups of modern human colonists into new geographic areas (Figure 6.2).

Multiple recent studies have surveyed more than 600,000 SNPs between people from diverse world populations. These SNPs are not from one specific place in the genome, but are spread over the entire genome and are from many different independent recombining segments. Three aspects of genetic variation indicate a serial founder effect from Africa. In the first, the amount of genetic variation decreases in those world populations that are sequentially farther and farther away from Africa. In the second, measures of the genetic distance between pairs of world populations, where one is an African population, indicate that populations at increasing distance from Africa show proportionately more genetic differences from the African population. Researchers call this pattern "isolation by distance," since populations farther and farther apart are less and less able to interbreed due to the increasing geographic distance between them. The third aspect lies in a measure of linkage disequilibrium. This fancy-sounding term expresses how much the genomes within a particular population

Figure 6.2: A serial founder effect whereby modern human populations successively lost DNA variation as they migrated and spread out of Africa. The letters represent different DNA variants in the population and the numbers give the counts of variants in each successive population.

have shuffled their many segments during the formation of the genomes of their offspring. Interestingly, Africans show the most shuffling of all human populations. Moving away from Africa, populations seem to have experienced less and less shuffling of genomic segments, described as greater and greater amounts of linkage disequilibrium. This pattern makes sense since the amount of shuffling of the genome depends on a population's age. If Africans represent the oldest human populations, they are expected to show the greatest amount of genomic shuffling. Likewise, if populations at greater and greater distance from Africa are also successively younger than African populations, then they are expected to show less shuffling.

Besides SNP genetic variation supporting a serial founding effect from an African origin, other types of genetic variation across the genome support the same scenario. One useful measure are regions of the genome where small motifs of DNA mutate as a unit, called microsatellites, to repeat themselves over and over again and create tracks of repeated DNA. For example, within a DNA sequence, the DNA motif "AGAT" might have been mutated to produce two repeated units of AGAT. (Look in the following DNA sequence to find this repeated motif: ACAACTTACTTAGATAGATAT CCCGAACAAT.) While the DNA might have mutated to have produced only a single repeat in one or several people within a population, it may have mutated in step-like fashion to produce three repeats in another group of people (ACAACTTACTTAGATAGATAGATATCCCGAACAAT). In other people in that population it might have grown to four or five repeats. Since microsatellites mutate much more quickly than SNPs, they are very useful for examining the differences in amounts of genetic variation within world populations, and for tracking population movements. There is an abundance of microsatellite regions spread out across our genome and located in many different segments. For each microsatellite region, researchers can determine the number of different repeat sizes found in people belonging to a particular population. In the example above, the population had as many as five different sizes of the microsatellite. When researchers have studied microsatellites from hundreds of different segments in the genome in many diverse world populations, and counted the numbers of repeats found at the different microsatellite regions, they have found a strong pattern of decreasing diversity, or numbers of repeat differences, within those populations farther and farther from Africa.[5]

A serial founder effect is also seen in human anatomical variation, and it too originates in Africa. Let's look at the shape of the cranium, our braincase. The cranium varies in shape within and between different world populations with shape differences traditionally thought to be the work of natural selection, perhaps to allow efficient thermoregulation in different climates. This idea has been formally tested in recent research with the finding that human cranial shape variation has by and large not been molded by natural selection, except in the case of peoples from very cold regions, where the nasal region, orbits, and cranial breadth were distinct. Instead, human variation in cranial shape appears to be a good indicator of population expansions since our evolutionary origin.[6] Interestingly, the shape of the human pelvis also seems to track past human movements, though the shape of our arms and legs does not.[7] Thus, as with genetic variation, the magnitude of head and hip shape variation observed within a population decreases in those populations that are sequentially farther and farther away from Africa.

Perhaps not too surprisingly, human tag-alongs like the parasites and bacteria that infect our body have not escaped human history but instead actually reflect our evolutionary origin and movements. *Helicobacter pylori*, a bacterium found in our stomachs, has been a faithful colonizer of our stomach ever since our species' origin, and seems to have traveled with us throughout our movement out of Africa and subsequent expansions. It also shows decreasing genetic variation in populations farther and farther away from Africa.[8]

This is also the case for the mosquito-borne human blood parasite *Plasmodium falciparum*.[9] All of these independent bits of genetic evidence, pointing in the same direction, amount to a massive consilience of evidence for our modern species' evolutionary origin in Africa and our subsequent expansions to colonize the world.

But where in Africa did we get our start? Africa is such a massive continent that the United States, China, India, and several additional smaller countries would fit within its boundaries, and so it requires us to be a bit more specific. Some genetic evidence, mostly from the mitochondrial genome, seems to point to an origin in East Africa, though other evidence coming from the nuclear genome points instead to South Africa. At present this question is an open one. As we shall discuss below, it is quite possible that our origins in Africa were more complex and geographically diffuse than the traditional idea of a single point origin. More fine-grained sampling of peoples from diverse African populations will help us determine this, and the underrepresentation of Africans in studies of genetic variation has certainly limited our ability to answer these questions.

Many studies to date that have detected a serial founder effect originating from Africa have studied the HGDP-CEPH collection of DNA samples. This panel of DNA samples assembled over the years by the Human Genetic Diversity Project and the Centre d'Etude du Polymorphisme Humain (CEPH) represents a collaboration between the great human evolutionary geneticist L. Luca Cavalli-Sforza from Stanford University and the 1980 Nobel Prize-winning French immunologist Jean B. Dausset, who founded the CEPH polymorphism center in 1984. The panel comprises DNA collected from 1,064 consenting individuals from fifty-one populations from sub-Saharan Africa and North Africa; Europe and the Middle East; South, Central, and East Asia; Oceania; and the Americas. This resource of DNA samples has been available to researchers since the early 2000s and represents a huge advancement for studies of human genetic variation. Previously, different laboratories would use very different sets of DNA samples from world populations, with some studies lacking many populations. Today, researchers can scrutinize the specific types of genetic variants that interest them, yet conduct their studies using the identical large and diverse set of DNA samples that other groups of researchers also use. For example, one lab may type hundreds of microsatellites in the HGDP-CEPH panel, another lab might study sets of SNPs in the panel, or yet another lab might scrutinize some other aspect of genome variation. The homogeneity in samples analyzed by different researchers removes the major concern among scientists that different findings between studies may simply be due to the fact that they have used different sets of DNA samples.

Although the development of the HGDP-CEPH panel is a major advance, it is still limited in its geographic coverage, with Africans being much under-represented. African samples in the panel come from only eight regional populations and there are very few samples of traditional hunter-gatherer groups, which we know are some of the oldest of all human groups.[10,11] This is very unfortunate considering the importance of Africans for studies of human origins, particularly studies that aim to locate the specific geographic region or groups of regions within Africa where modern humans evolved. It is also important to have more diverse samples in order to understand the genetic bases for regional human diseases in Africa that could ideally lead to treatments and cures.

Another important concern is the bias that exists in many of the genetic variants that we have discovered. Many of the genetic variants between different peoples and populations that appear in our databases were discovered in initial studies of very restricted samples. Historically, studies focused on European populations in search of clinical applications of the detected genetic variants. These arrays of hundreds of thousands to millions of DNA variants were then assembled into a detection kit—a DNA microarray plate or DNA-chip—onto which new DNA samples were placed to see if they contain the known variants (Figure 6.3). Thus when DNA-chips are used to screen for DNA variants they may show unrealistically inflated levels of variation in certain populations yet much lower levels in others. This bias is called "ascertainment bias." For the reconstruction of the evolutionary histories of populations, such biases have the real potential to lead to spurious results and any researcher interested in using DNA-chips must always bear this in mind. For example, the population geneticist Rasmus Nielsen and colleagues at the University of Copenhagen have explicitly studied the levels of ascertainment bias associated with using a popular DNA-chip, and found that estimates of the degree of population differentiation and the genetic distance between populations can sometimes be seriously compromised.[12] Nevertheless, the multiple studies that support a serial founder effect appear to have had minimal bias, and this is especially so for the microsatellite evidence which is thought to be mostly immune to ascertainment bias because its mutation rate is very rapid.

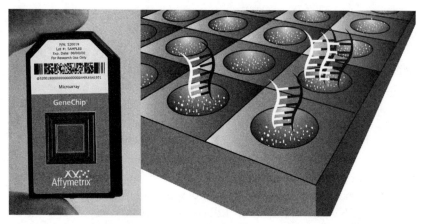

Figure 6.3. On the left, a microarray plate or DNA-chip, used in the detection of known DNA variants. (Courtesy of Affymetrix, Inc., Santa Clara, CA, USA.) The image on the right shows how DNA is placed into each of thousands of tiny wells inside the DNA chip, with each well designed to detect a specific DNA variant.

Today, due to radical technological advancements as well as to the general understanding among researchers of the problems of bias, there is a movement away from microarray studies to large studies that fully sequence genomes determining all their DNA bases, like the 1000 Genomes Project initiated in 2008 (see chapter 7 for more details on this project). Full sequencing of genomes allows the discovery of new DNA variants in the genomes of people around the world, many of which we would not otherwise have known about. A recent study led by evolutionary geneticist Sarah Tishkoff at the University of Pennsylvania sequenced fifteen genomes of Africans from three different hunter-gatherer populations: Pygmies from Cameroon and the Hadza and Sandawe from Tanzania.[11] As a result, the researchers were able to contribute over 5.5 million previously unknown DNA variants to the public database. Together, these make up a whopping 41% of today's database of human DNA variation, but this also shows how limited our understanding of human diversity has been in its underrepresentation of Africans. Developing a database comprising the full genomes of thousands of people from around the world will be a huge advancement for evolutionary studies.

HOW QUICKLY DID *HOMO SAPIENS* EMERGE?

Several elements have predisposed scientists to the idea that our evolutionary origins took place as a discrete event, focused both within a small region in Africa and a small window of time. Human skulls with anatomically modern features—the large, rounded, and globular braincase; the forehead that rises vertically; a flat face tucked beneath the braincase—are known from the Ethiopian sites of Herto Bouri and Omo Kibish and are dated by absolute radiometric methods to approximately 160,000 and 195,000 years ago, respectively.[13,14] Yet, between those sites and from as far back as 700,000 years ago, we have found a series of fossils from regionally disparate locations in Africa (North, South, and East) that appear to gradually change from an archaic human anatomy to a more modern human anatomy. In aggregate, these skulls come from geographically disparate regions including North Africa, East Africa, South Africa and even Southwest Asia (Israel), and combine modern human features with more archaic features in different ways in different fossils (Figure 6.4). This series

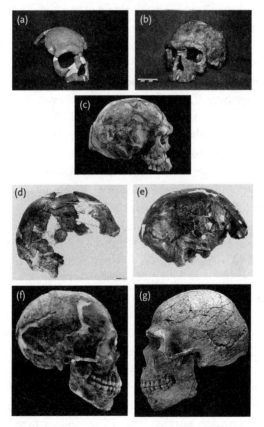

Figure 6.4: A "line-up" of early modern human fossils from different geographic regions in Africa and Southwest Asia (Israel). These fossils and others appear to indicate that *H. sapiens* evolved over a broad geography including these areas and perhaps additional regions of Africa. (a) Florisbad (South Africa); (b) Jebel Irhoud (Morocco); (c) Herto (Ethiopia); (d) Omo 1 (Ethiopia); (e) Omo 2 (Ethiopia); (f) Qafzeh (Israel); (g) Skhul 5 (Israel). (Credits: Qafzeh 9, Skhul 5 and Herto from David Brill; Florisbad and Jebel Irhoud from Günter Braüer; Omo I and 2 from Michael Day).

of fossils has led the paleoanthropologist Günter Bräuer at the University of Hamburg to hypothesize that modern humans evolved in a gradual and mosaic-like fashion from an early archaic grade of *Homo sapiens* (between 600,000 and 300,000 years ago) to a late archaic grade (between 300,000 and 200,000 years ago), and then to an anatomically modern grade *of Homo sapiens* (around 200,000 years ago).[15,16] This mosaic pattern of evolution is underscored when we compare the two skulls found at Omo Kibish known as Omo I and Omo II that are dated to approximately the same time. The Omo I cranium has very modern anatomical features, yet Omo II combines modern features with several robust and archaic features; for example, the back of the braincase is angulated, and a bony keel runs from front to back along the top of the skull.[15] It is not known if they come from a single very variable population or from two different populations,[17] but either way they appear to support Bräuer's gradual and mosaic model of the evolution of our modern species in Africa.

A gradual scenario like Bräuer proposes is very different from a scenario in which our species evolved within a short time frame and more recently, as has often been alluded to or explicitly hypothesized in "out of Africa" models of modern human origins. The idea of a discrete temporal event for the origin of our species has also been strongly influenced by the analyses of mitochondrial DNA, the first DNA system analyzed in diverse human populations.[1,18] The human mitochondrial tree coalesces back to approximately 200,000 years ago. Many researchers have felt the mitochondrial tree indicates a population bottleneck at the time of our species' origin in Africa, in which the population ancestral to modern humans was significantly reduced to several thousand people within a single geographic region in Africa (see Weaver, 2012).[19] This location has often been surmised to be East Africa, because the Omo I and Herto fossils date from about the time of mitochondrial coalescence, and also because East African hunter-gatherer groups are believed to have some of the oldest mitochondrial lineages.

However, in 1999 when Jody Hey and I determined DNA sequences of the *PDHA1* gene in diverse human populations, it led us and other researchers to surmise that our species' origin might not have been as simple as the mitochondrial diversity suggested.[20] At the time, *PDHA1* was one of the few nuclear genes that had been studied in diverse populations, but it represented a growing number of nuclear genes where DNA variation seemed to be incongruent in some important ways with mitochondrial DNA variation. Although most nuclear genes favored the view that modern humans' origins were in Africa (as did *PDHA1*), the ages of the gene trees built for these genes were almost in every case very old compared to the mitochondrial tree. For example, *PDHA1* variation coalesced back to 1.7 million years ago—almost eight times older than the mitochondrial DNA tree. Other nuclear gene trees were almost as old or older. If there had been a strong population bottleneck at the time of our species' origin, then we would not expect these nuclear genes to coalesce so deeply in the past.

The big questions were shifted. Were the old coalescence dates for nuclear genes a fluke? If they were not, what could this tell us about the evolution of modern humans, and what would this indicate about the likelihood of a severe bottleneck when our species evolved?

Indeed, the nuclear genome offers numerous opportunities in its independent genomic segments to test these ideas. One way to do this is to build separate trees for independent genes or segments of the genome, and determine the age at which each of these trees coalesces back in time to a single common ancestral DNA copy. The time to coalescence of independent segments of the genome is expected to vary widely, since the coalescent process at most genomic segments is largely under the influence of random evolutionary processes such as genetic drift. However, it is the average age of coalescence across many genes, as well as the degree of spread of coalescence dates into the past, that will be important in determining whether there was a severe bottleneck at the time of our species origin. In 2012, Michael Blum at the Université Joseph Fourier in France studied over sixty nuclear genomic segments in large sets of individuals from several African populations including the San Bushmen and Biaka Pygmy hunter-gatherers. The segments came from different regions of the genome, and none of the regions contained any genes. Regions with genes should be avoided for these types of studies, since these regions can yield gene trees of unusual shapes due to the effects of natural selection. The analysis was restricted to African populations because if there had been a bottleneck at the time of our origin, it should be most evident in the genetic variation of African populations. We also know that non-African populations experienced a population bottleneck during the "out of Africa" expansion (about 50,000 years ago), so including these populations in the analyses could interfere with our ability to detect an earlier bottleneck at the time of our species' origin.

Blum's study determined that the set of autosomal regions (i.e., all chromosomes except the sex chromosomes) had an average coalescence date of over 1.5 million years and the X chromosomal gene regions they looked at had an average coalescence date of over 1.0 million years ago. For all sixty regions, the spread of coalescence dates was enormous, from 380,000 years to as long ago as 2.4 million years. These ages are hard to reconcile with the 200,000-year mtDNA coalescent date and the large spread of dates seems hard to square with the idea of a severe population bottleneck at the time of our species' origin.[21,22] In addition, after a bottleneck the differences in population measures of DNA diversity are theoretically expected to vary widely across the different regions; however, these statistics are instead more uniform from region to region, which Blum interpreted as a rejection of the idea of a bottleneck.[23] In another study in 2011, researchers came to the same conclusion using different techniques and examining larger sets of DNA variants spread across the genome.[24] In fact, this analysis estimated that African population sizes were actually slightly increased during the time of the purported speciation bottleneck 100,000 to 200,000 years ago.

A rather strong clue that our modern species did not evolve within a single distinct region of Africa can be seen in the enormous spread (over 2.0 million years) in the coalescent dates of autosomal and X chromosomal gene regions. Theoretical work by population geneticist John Wakeley at Harvard University has shown that when a new species evolves in the midst of multiple geographically separate subpopulations, with migration and interbreeding connecting them, different regions in the genome will have segments with vastly different times to their coalescence. These new models are described as metapopulation models,[25-27] represented in Figure 6.5. A metapopulation scenario for modern human origins describes multiple ancestral subpopulations located in close but different geographic regions within Africa, with subpopulations connected by webs of gene flow as individuals migrated to varying extents among subpopulations. The idea may very well provide a biologically more realistic explanation for our species' origin, though one that is decidedly more complex than the story based solely on mitochondrial DNA. The scenario is also a dynamic one, recognizing that in the real world subpopulations can and do experience unique changes through time. Some may undergo temporary contractions in size, others may temporarily expand or go entirely extinct, while other subpopulations may form new subpopulations through the colonization of a nearby region. This new metapopulation scenario for the origin of modern humans is compared to a mitochondrial-based scenario in Figure 6.6.

We are fortunate to have many ancient autosomal gene regions within our genome. Together they provide a window through which we can peer back at events in our species' history that occurred before the recent times when individuals from the sites of Omo and Herto (in Ethiopia) lived. They allow us to ask more pointedly if multiple human subpopulations in Africa, connected by a genetic web, evolved into our modern species. Even early studies of PDHA1 and other genes hinted that there might have been multiple subpopulations that gave rise to the modern human gene pool, though none could provide any degree of certainty on where these different

Geographic Subpopulations

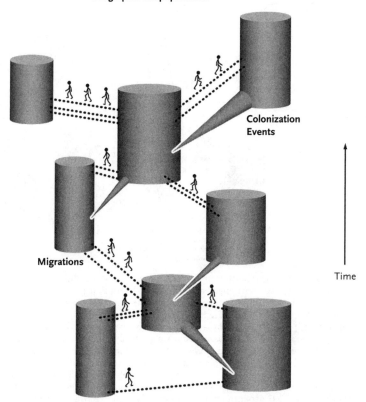

Colonization
Events

Migrations

Time

Figure 6.5: A hypothetical example of a metapopulation model of human origins. Vertical cylinders represent different geographic subpopulations ancestral to modern humans within Africa. Dark gray connections represent colonization events. Thin dashed lines represent migrations of people between groups with subsequent interbreeding and flow of genes between subpopualtions.

subpopulations were located.[20] For example, *PDHA1* showed the possibility of different subpopulations existing in the timeframe of 200,000–400,000 years ago. Other nuclear gene regions analyzed by the population geneticist Michael Hammer and colleagues at the University of Arizona also showed evidence of multiple ancient subpopulations contributing to the evolution of our species.[27] In the future we look forward to further refining the metapopulation model for the origins of modern humans. If the model is correct, we will want to know exactly where these multiple subpopulations were located. Which, if any, of the African fossils that we have discovered so far were our ancestors? Did each subpopulation contribute equally to our modern genome, or did some play more important roles than others? We should come closer and closer to answering these questions as

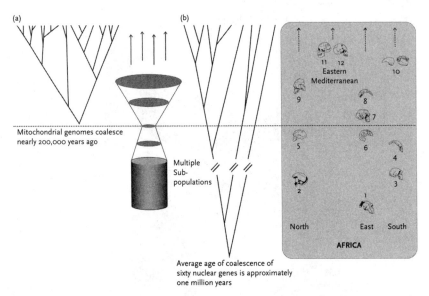

Figure 6.6: Two models of evolution of modern humans in Africa. The illustration on the left (a) posits a population bottleneck at the time of speciation of modern humans in a single population in Africa. The mitochondrial gene tree, which coalesces back to nearly 200,000 years ago, has been one of the main pieces of evidence for this model. The illustration on the right (b) posits that multiple subpopulations in geographically different regions in Africa over several hundred thousands of years gradually evolved into modern humans. Two aspects of gene trees from many independent nuclear genes support this model: the great spread of coalescent dates for genes in the nuclear genome, and second, that many gene trees are disproportionately ancient. Fossils on right: (1) Bodo, Ethiopia; (2) Sale, Morocco; (3) Kabwe, Zambia; (4) Florisbad, South Africa; (5) Jebel Irhoud, Morocco; (6) Omo 1, Ethiopa; (7) Herto, Ethiopia; (8) Aduma, Ethiopia; (9) Dar es Soltane, Morocco; (10) Klasies River Mouth, South Africa; (11) Skhul 5, Israel; (12) Qafzeh 9, Israel.

we study the genomes of more and more diverse Africans. Furthermore, if we are someday able to extract the ancient DNA and then determine the ancient genomes from the different fossil contenders for early ancestors of modern humans in Africa and Southwest Asia (such as those shown in Figure 6.4), this will be particularly revealing.

DID OUR EARLY AFRICAN ANCESTORS INTERBREED WITH ARCHAIC HOMININS?

While entire ancient genomes have been extracted from fossils in Europe and Asia (see chapter 8), to date no ancient DNA has been extracted from African fossils. We are not even sure if DNA still exists in these fossils, since the environmental conditions the African fossils endured (often open air and hot and wet environments) were very different than in Eurasia, where most ancient DNA comes from fossils found at cave sites and in regions where colder environmental conditions prevailed for many millennia. Even without ancient DNA from these fossils, genome-wide DNA evidence from Africans alive today points to the likelihood that they interbred in the past (to a small degree) with distant archaic cousins not on the direct lineage leading to modern humans. In separate research studies, Sarah Tishkoff and Michael Hammer have found genomic evidence that indicates that African hunter-gatherer populations, including Pygmies from central Africa, San Bushmen from southern Africa, and the Hadza and Sandawe from Tanzania, contain up to 2% archaic DNA in their genomes.[11,28] These unusual genomic regions look much more ancient than other parts of their genomes. For one, they show minimal signs of having been shuffled by recombination, suggesting they were imported into African genomes through interbreeding with long-lost isolated evolutionary cousins. The genomic segments are also estimated to be very old, appearing to have come from either ancient modern human populations or *bona fide* ancient hominin species that split from the human lineage at different times in the past, ranging from 375,000 years ago to as long ago as 1.5 million years ago.

We are not sure who these archaic cousins were, but they may have been members of the group Günter Braüer has described as the earliest grade on a continuum directly leading to *Homo sapiens* in Africa, evolving as early as 700,000 years ago. The genomic studies estimate that interbreeding occurred at multiple times in the past and, surprisingly, until very recently. Tishkoff's group found evidence that a major interbreeding event occurred in the time before African hunter-gatherer groups differentiated, probably in the time range of 100,000 to 150,000 years ago. Hammer's group estimated that interbreeding continued until around 35,000 years ago, when many populations in Africa already had modern features. This indicates that some of our most archaic relatives in Africa probably survived in small isolated populations until very recently. For example, a complete skeleton from Nazlet Khater in Egypt and a braincase from Hofmeyr in South Africa, both dating to approximately 35,000 years ago, show robust and possibly archaic features.[29] Such fossils could represent lingering, and perhaps semi-isolated, archaic human populations with whom anatomically modern humans in Africa still interbred.

To confirm that interbreeding between archaic hominins and modern humans took place in Africa, and to find out who these archaic hominins might have been, it will be necessary to extract DNA of sufficient quality from African fossils so that their genomes can be assembled. We can then do a direct comparison to see if ancient regions detected by Tishkoff and Hammer in modern African genomes match up with segments of the genomes extracted from African fossils. This will give us direct proof. We will also want to know if there were any genes with important functions passed from archaic hominins into the early African gene pool through these ancient matings, and if these regions contributed to us becoming human.

SQUEEZING THROUGH A BOTTLENECK

The serial founder effect suggests that as modern humans spread from Africa, each new geographic region that was colonized and each new population that was founded represented a subset of the genetic variation of its parent population. Although this generally appears as a smooth process, there were certain times when more dramatic drops in genetic variation occurred. One of these events occurred when modern humans expanded out from Africa. A population bottleneck out of Africa is supported by analyses of the mitochondrial genome. And the bottleneck has in the last several years been supported by studies of many regions of the nuclear genome, which have also allowed an assessment of its timing and severity. The results from multiple large studies suggest that the bottleneck occurred between 31,000 and 55,000 years ago,[30] with the earliest times (around 55,000) being supported in studies using improved methods.[31] The oldest part of this range is consistent with the age, approximately 42,000 years, of some of the first anatomically modern human skulls found in southeastern Europe at the Romanian site of Peçstera cu Oase ("The Cave of Bones").[32] The bottleneck also appears to have been quite severe, with evidence from most studies indicating an 80% to 90% reduction in population size. If we use the guide of a one to three ratio between effective population size to census size,[33,34] admittedly a rough guide, this works out to be a reduction from an African population size of 30,000 people down to around 4,500. As we have mentioned earlier, the conversion from effective size into census size is complex, so while the exact population sizes may indeed be underestimated, the severity of the collapse in population size as we spread out of Africa is likely to be quite accurate.

Although a consensus among studies is emerging regarding the time frame and severity of this population bottleneck, many important questions remain. Most studies assume that the start of the out of Africa bottleneck indicates the time of the split between Eurasian and African populations, and the end of migration and interbreeding between them. However, this is questionable. Analysis of nuclear genomes within the last several years has found evidence for substantial levels of gene flow indicating continued interbreeding between African populations and Eurasians after their initial split and continuing to as recently as 20,000 to 15,000 years.[24,31,35] In the future it will be interesting to pinpoint which Eurasian and which African populations remained in contact over tens of thousands of years and continued exchanging migrants and mates. For example, in Africa, was it primarily western, northern, or eastern Africans that remained in reproductive contact with Eurasians? Which Eurasian populations were involved? Did the populations involved in continued reproductive exchange over tens of thousands of years shift over time, and, besides genes being exchanged, what were the cultural effects of continued contact?

Another important question is whether there was more than one population bottleneck, the possible consequence of more than one dispersal event from Africa. At least one recent genomic study has pointed to two dispersal events, with a possible human dispersal into Southeast Asia preceding the dispersal into Europe and east Asia by perhaps tens of thousands of years.[36] Such a scenario might fit in well with evidence from archeology and fossils that indicates that modern humans moved into southern Asia and into Southeast Asia prior to colonizing Europe and east Asia.[37]

An "out of Africa" population bottleneck caused a large reduction in amounts of DNA variants in the genomes of present-day European and Asian peoples. One study, based on genome-wide DNA sequence data, found that Eurasian peoples show about 30% fewer genetic differences than do African peoples.[38] If we compare the full genomes between individuals as far apart as Asia and Europe, we find many fewer DNA differences between them than between two Bushmen of the Kalahari Desert of southern Africa.[39] Besides having reduced genetic variation compared to Africans, populations outside Africa also have significantly more (approximately 25% more) damaging DNA variants thought to have slightly deleterious effects compared to Africans.[40] This is likely due to reduced negative selection during the bottleneck when the size of the population expanding out from Africa was reduced. Recent genomic research has found that such damaging variants, while very rare in the population, could well underlie complex diseases within Eurasian populations.[41]

Two aspects of human demographic history can explain the increase in damaging DNA variants in Eurasians. As described in chapter 4, a drop in population size like the out of Africa bottleneck will alter the evolutionary forces acting in the population. In the bottlenecked population, the random evolutionary force of genetic drift dominates over negative selection, the evolutionary force that normally would weed out damaging DNA variants. The damaging DNA variants in the bottlenecked population will therefore persist in the population, simply because negative selection has reduced power to discriminate between good and bad variants. The second factor is the population increase in non-Africans after the bottleneck. Although the mutation process occurs continuously in evolution, when a population is expanding in size, the new mutations that accrue will mainly be at those sites that cause functional changes in the gene. (Remember, because of the way that DNA codes for proteins, 75% of sites, if changed, will cause a functional change.) Thus, population growth alone after an out of Africa population bottleneck will increase damaging variants in genes. You might reason that because world population size is now downright huge in non-Africans, as it is for the entire world human population, purifying selection should be strong enough to remove these damaging variants. This is true—but in the context of evolutionary time, human populations have only become very large relatively recently. The human population is thus experiencing a time lag between when our populations began to grow and when negative selection will fully kick in and work full steam at removing damaging variants. Because world populations are so huge today, we can expect the prevalence of damaging variants in human genes to decline in our species' future.

After spreading out of Africa, modern humans dispersed to colonize Eurasia and eventually most of the world. Recent studies have discovered that the very process of rapid spatial expansion into new geographic regions can have damaging genetic effects. For example, a detrimental genetic variant that is rare within a population before it spreads can later become very common in populations at the end of an expansion. This process is known as gene surfing and its consequences are only beginning to be studied (Figure 6.7). More precisely, as populations expand into new territories, DNA variants that exist in the genomes of people at the very front of the wave of advance experience strong random genetic drift because of the drastically reduced population sizes present at the front. Because genetic drift is intensified at the leading edge, negative selection is less effective at weeding out any deleterious variants that may be present. Therefore, gene surfing could have inadvertently spread some of the DNA variants in Europeans that cause diseases. Some examples of diseases that might have become common through surfing include hemochromatosis, in which individuals store excess iron in their bodies, and cystic fibrosis, in which people secrete thick mucus into their lungs. In Europe, DNA variants underlying these diseases become increasingly common in populations as we move along a trajectory from southwest Europe (where they are rare) to northeast Europe (where they are more common). These disease variants could have surfed on an ancient expansion of peoples from the Near East.[42]

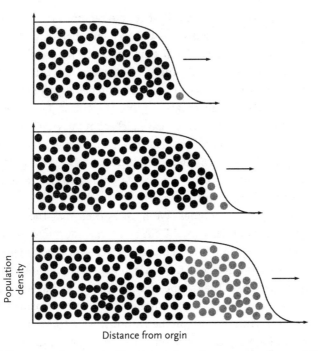

Population density

Distance from orgin

Figure 6.7: From top to bottom, the illustrations show the expansion of a population in the direction of the arrow. The top illustration shows a new mutation (gray dot) arising on the wave front of the advancing population. Low-frequency mutations that appear on the wave front have a larger chance to "surf" to higher frequencies in the descendant populations (bottom two illustrations) than do low-frequency mutations within the larger populations inhabiting already occupied geographic regions (area with black dots).

We are still unsure to what degree human expansions caused gene surfing. But we do know that human populations expanded into new geographic areas multiple times in our past. There was the initial expansion out of Africa into Eurasia about 50,000 years ago, and expansions that occurred at the end of the last glacial maximum about 18,000 years ago. Finally, about 10,000 years ago, people carrying agriculture from the Near East expanded into Europe, replacing and partially interbreeding with hunter-gatherer peoples already living in these regions. A challenge for the future will be to figure out more precisely where and when these multiple expansions occurred and how they spread different genes, some of which cause diseases, across Europe and Asia and other parts of the world.

FITTING THE PUZZLE PIECES TOGETHER

As students of our evolutionary past, we must realize that our questions cannot and will not be answered by sequencing genomes alone. Assembling the puzzle of human evolution will of course require fitting together the diverse pieces of genetic evidence—the vast number of DNA segments making up nuclear genomes, the ancient genomes we are able to extract from fossils, the mitochondrial genome. But more than this, we will need to figure out how fossil skulls and bones and archeological remains (tools, ornaments, and other artifacts) also fit into the puzzle. Only then can we assemble a coherent and comprehensive picture of our past.

Considering genomes alone, there are several challenges that researchers of modern human origins will need to continue working on. One of them is to reconcile the information from uniparental genetic systems—mitochondrial genome and Y chromosomes—with what we are now learning from complete nuclear genomes. We have learned that the very shallow coalescences of the mitochondrial DNA and Y chromosomal trees are very limited in what they can tell us about our species' origin. The much older ages of gene regions across our nuclear genome are providing us with evidence that the processes that shaped our emergence as a species were considerably more complex than mitochondrial or Y chromosome evidence alone can reveal. A challenge for the future will be to learn the ways in which these different genetic systems can shed light on different aspects of our past. Because mitochondrial DNA and Y chromosomes are inherited either maternally or paternally, they will likely play important roles in helping us to understand the special dynamics that females and males played in our past. For example, Lounès Chikhi and Rita Rasteiro at the Gulbenkian Institute of Science in Portugal studied the shift from hunter-gatherer lifestyles to agricultural lifestyles in Europe that occurred approximately 10,000 years ago, using both mtDNA and Y chromosomes. They found differences in patterns of genetic variation between the two systems that suggest a shift toward more sedentary lifestyles, but also an increase in female migration due to a new cultural practice where wives moved to live on their husbands' family farms.[43]

The faster mutation rate of the mitochondrial genome compared to the nuclear genome might make it especially useful for gaining insights into the most recent events in our evolutionary past, although we always need to be careful not to overinterpret its importance, since it is only a single piece of genetic evidence. As Chikhi has noted, using mitochondrial DNA variation alone to investigate ancient migrations can often amount to mere storytelling—"mythochondrial" DNA stories, as he calls them![44]

Due to great progress in laboratory methods and technology, it is becoming relatively straightforward to extract ancient DNA from fossils and assemble full genomes from our ancestors. We are at the point where, if DNA is still preserved within a fossil, it will be possible to obtain and extract it and analyze it. As more and more ancient genomes become available to us, such as from Neandertals and other archaic ancestors, we can explore better how we are related to them. There are many interesting questions to examine. When did we last share a common ancestor with them? When modern humans spread out of Africa, did we meet Neandertals and other archaic hominins who were already living in Europe and Asia? If we did, what were our interactions like? For example, did we interbreed with them, and if so to what extent did we interbreed (see chapter 8)? Did our interbreeding result in modern humans acquiring new and important functional regions in our genome? Any full explanation of modern human origins will have to take into account any information we glean through the analyses of ancient genomes.

For researchers in all fields of human evolutionary studies, the continuing challenge is to figure out how all the disparate pieces of evidence of our past can be reconciled to form a complete single picture. For more than a few researchers, fossil and archeological evidence do not always seem to fit easily within the timeframe of our evolutionary past as estimated in genomic studies. Some researchers have proposed that if the timeframe from genomic evidence could be readjusted—essentially turning back the clock—it would help solve any incongruence between stones, bones, and genomes.

The rate at which mutations tick away through evolutionary time is used to estimate the times at which evolutionary events occurred, such as when chimpanzee and human lineages separated. As mentioned previously, this mutation rate calibrated with the radiometric date of a fossil is known as the phylogenetic mutation rate. Two recent findings have caused us to take a new look at the accuracy of the mutation rates we use. First, great apes and humans appear to have experienced a significant evolutionary slowdown in their mutation rates since splitting from other primates.[45] Second, with advancements in sequencing technologies, today's well-equipped laboratories can sequence entire individual genomes of three billion nucleotides in a short period of time, and therefore the number of mutations between parents and their children can be determined directly and a new type of mutation rate, known as a pedigree or *de novo* mutation rate, can be obtained. Strikingly, *de novo* mutation rates are extremely slow, and appear to be almost half as slow as the rate traditionally used to estimate events in human evolution. What does this mean for the timeline of events for human evolution? Are they twice as old as we previously thought?

At present it is still unclear if mutation rates between present-day parents and offspring can be generalized back over hundreds of thousands or even millions of years in order to estimate the timeline of human evolution.

Indeed, some recent research using ancient DNA extracted from fossils with radiometrically determined ages appear to support the traditional phylogenetic mutation rates. Nevertheless, in a provocative 2012 paper, Aylwyn Scally and Richard Durbin at the Sanger Institute for genome research in Britain have described the implications of the *de novo* mutation rate for the time scale of human evolution, and how many apparent incongruences between genomes and fossil and archeological evidence disappear using this new rate.[45]

Using phylogenetic mutation rates on full genomes, the dispersal out of Africa is estimated to have occurred about 50,000 years ago. As mentioned above, this poses a problem for evidence that suggests an early migration of humans into Southeast Asia. For example, the analysis of the Australian Aborigine genome estimated that the ancestors of these people split from Eurasians as long ago as 75,000 years ago. At the same time, some archeologists like Michael Petraglia at the University of Oxford and his collaborators have found artifacts that they believe indicate that modern humans dispersed out of Africa and into Yemen, Arabia, and India at very early times—over 85,000 years ago and perhaps even earlier.[37] Studies of artifacts from archeological sites in these regions reveal similarities to artifacts in Africa. Although these people didn't possess the exact cultural package of modernity seen at more recent European sites—ornate stone toolkits, built structures, diverse body ornaments, and rock, ivory, and bone art—they could still have been part of the initial expansions of modern humans out of Africa.

Such an early move out of Africa by modern humans challenges the belief held by other archeologists, like Richard Klein at Stanford University, that modern humans only successfully expanded from Africa into Eurasia only once they crossed a threshold of cultural modernity, the so-called "human revolution" hypothesis.[46] The slower *de novo* mutation rate places the first modern human dispersal from Africa twice as far in the past, about 90,000 to 130,000 years ago. Such an early estimate might indicate that the artifacts discovered by Michael Petraglia's group are indeed part of the initial expansion of modern humans from Africa. In fact, an earlier timeframe for modern humans leaving Africa might also indicate that the people that inhabited the cave sites at Skhul and Qafzeh in the Levantine eastern Mediterranean from 100,000 to 130,000 years ago were also part of the initial expansion from Africa. Researchers have traditionally figured that habitation of modern humans at these cave sites represented an ephemeral occupation permitted by a phase of climate warming. But if the *de novo* mutation rate prove correct, then the Skhul and Qafzeh peoples could conceivably be ancestors to all non-African peoples. Of course, the ancient genomes from these possible ancestors, if we can get them, could tell us for sure.

We all recognize that puzzles become harder to solve the more pieces they contain. The genome is like a giant puzzle with its pieces consisting of many thousands of recombining genomic segments. Today, the field of human evolution has moved well beyond studies of single genes to analyzing the many thousands of unlinked genomic segments making up our genome. While it is more challenging to assemble the abundant pieces together in a genome, it is extremely important to realize that having many different recombining segments of the genome, each of which can have potentially different evolutionary histories, is hugely beneficial and makes it easier to reconstruct our evolutionary past. A single genomic segment can show us a single history, but we can never know if this history accurately reflects the evolutionary past of our species. As we saw in our quest to find our closest evolutionary cousin among the apes, any single genomic region can be absolutely misleading. Likewise, we cannot accurately know about our modern species' origins, migrations, bottlenecks, expansions, and interbreeding with archaic hominins (all population-level processes) until we find the same history corroborated by more and more independent segments of our genomes.

In fact, as more and more independent pieces of our genome are found to share the same history, we can be more and more certain that we have uncovered the actual evolutionary history of our species. On the other hand, those pieces of our genomic puzzle that stand out from the common pattern, even if just slightly, may give us information that these regions include genes or genomic regions that played important functional roles in us becoming humans—for example in the origins of our language and cognitive abilities. A lot of our success in accurately understanding our evolutionary history will depend on how accurately we can discern the common pattern across the genome, and how well we can tease out the small fraction of genomic regions that do not fit the common pattern.

CHAPTER 7

The Ongoing Evolutionary Journey

Humans today inhabit almost all land regions on Earth. It has been this way for thousands of years. Our colonizing adventure began about 50,000–60,000 years ago when humans in our modern anatomical form left Africa and migrated into Eurasia. As modern humans moved into and began to live in new regions of the world, they encountered very different environments that subjected them to new selective stresses. Migrating populations faced differences in temperatures, humidity, foods, disease-causing pathogens, intensity of ultraviolet light, altitude, and other challenges. These new selective stresses influenced survivability and the ability to successfully reproduce, and so we believe that natural selection has acted on populations over thousands of years in these new environments.

It is easy to see some of the ways that people vary in their appearance from region to region—in average body shape or height, differences in skin coloration, hair color and texture, and eye shape, among other features. Some features are not easily observable but involve inner physiological adjustments to enable different diets, to maintain sodium homeostasis and regulate body temperature, or to help ward off infections due to pathogens unique to different geographic regions.

To what extent are these physical and physiological features grounded in adaptations in our genomes? More specifically, to what extent have our genomes been shaped by the evolutionary force of natural selection, promoting the spread of favorable genetic variants to help people survive and reproduce amid the stresses imposed by living in new environments? Can we discover new genetic adaptations by comparing the genomes of people from different human populations? Can we identify the types of genes that have been most affected by natural selection?

Analyses of the adaptations of regional human populations are fundamentally different from studies of our species-wide adaptations. Population analyses do not require comparing genes and genomes between humans and other species, such as the chimpanzee. Instead, finding these adaptations requires scientists to compare genes or genomes between people living in different regional populations. Already, genomic studies have compiled large data sets of the genetic differences among diverse people. These projects are growing in size every day at a very rapid pace. The 1000 Genomes Project is one such coordinated effort,[1] among other smaller ones, that aims to sequence the complete genomes of thousands of geographically diverse individuals. These large amounts of genomic information will enable an even more comprehensive and fine-tuned look at how people came to thrive under differing environmental conditions.

In this chapter, I discuss some of the important and sometimes unexpected recent findings about population-specific adaptations from genome and gene studies, especially adaptations due to diet, levels of ultraviolet light intensity, disease agents, and high altitudes. It is not always so straightforward to recognize signatures of adaptation in the genomic data. One bugbear is how demographic dynamics in past human populations like population reduction or expansion can create a false signature of selection or in some cases obscure the true signal. We have also come to learn that some of our ideas about the evolutionary process of adaptation are rather simplistic. Previously, we had expected to find a type of signature when an adaptation is due to a single gene, where a newly arising beneficial mutation is spread by natural selection to every member of the population immediately after the mutation appears in the population. But such clearcut or distinct signatures of adaptation seem to be rare in our genome. Many adaptations in human populations could have left softer signatures within our genome, requiring new, qualitatively different, and more sensitive approaches to detect.

A HITCHHIKER'S GUIDE TO NATURAL SELECTION

In a classic paper published in 1974, the late John Maynard Smith and John Haigh at the University of Sussex described how natural selection acts upon a new beneficial DNA variant that arises within a population.[2] Remember that mutations regularly occur over evolutionary time. Most mutations are negative and are eliminated either through death or because their bearer has no children, or are neutral and have no effect on whether one lives, dies, or reproduces. It's rare that a beneficial mutation arises. But when it does, Maynard Smith and Haigh suggested that this mutation will spread rapidly through the population for which it is beneficial, until most or all of the individuals in the population possess it. Natural selection that acts to promote a beneficial mutation is known as positive selection. The key point of Maynard Smith and Haigh's paper was to demonstrate the effects that the spread of the beneficial mutation would exert on other mutations located near the beneficial mutation. They called this a selective sweep.

In Figure 7.1, we see the DNA sequences for seven different people in a population. In this illustration only eighteen bases of DNA are being compared between people for simplicity. We see three phases of a selective sweep: before the selective sweep, the completion of the sweep, and a short time after the selective sweep. In the first phase, note that the DNA bases of each individual are mostly identical, though at four sites the bases differ (at sites 2, 6, 13, and 16). The first three of these sites (2, 6, and 13) have no functional effect, or are neutral. There is one site (site 16), however, at which the DNA variant in the sequence copy in individual 3 is beneficial (indicated by a star).

At the completion of the selective sweep (second phase), we see that the beneficial DNA variant (starred) has spread to all individuals in the population through the action of positive selection. For this to have occurred, several generations must have elapsed, with the most rapid reproducers bearing the beneficial variant. Their children will then have inherited the beneficial variant, and so on until the point that all people in the population carry it. Aside from the spread of the beneficial variant, there are two other very important aspects of the second phase to observe. The first is that the three neutral variants that were on the same sequence that bore the beneficial variant are now found (like the beneficial mutation) in all individuals, even though they themselves are neutral. This so-called genetic hitchhiking of neutral mutations occurs because all the DNA sites near to each other on the same chromosome, which form a *linkage group*, are inherited as a single unit. In essence, they are "glued" together. The second point is that all DNA variants present in the first phase that were not on the third DNA copy have disappeared from the population by the end of the sweep. These mutations have been "swept" away as the copy bearing the beneficial variant has replaced all other sequences.

The third phase follows a short time after the completed sweep. In this phase, the mutational process has already started to produce new DNA variants in the DNA sequence copies from different people. Since only a

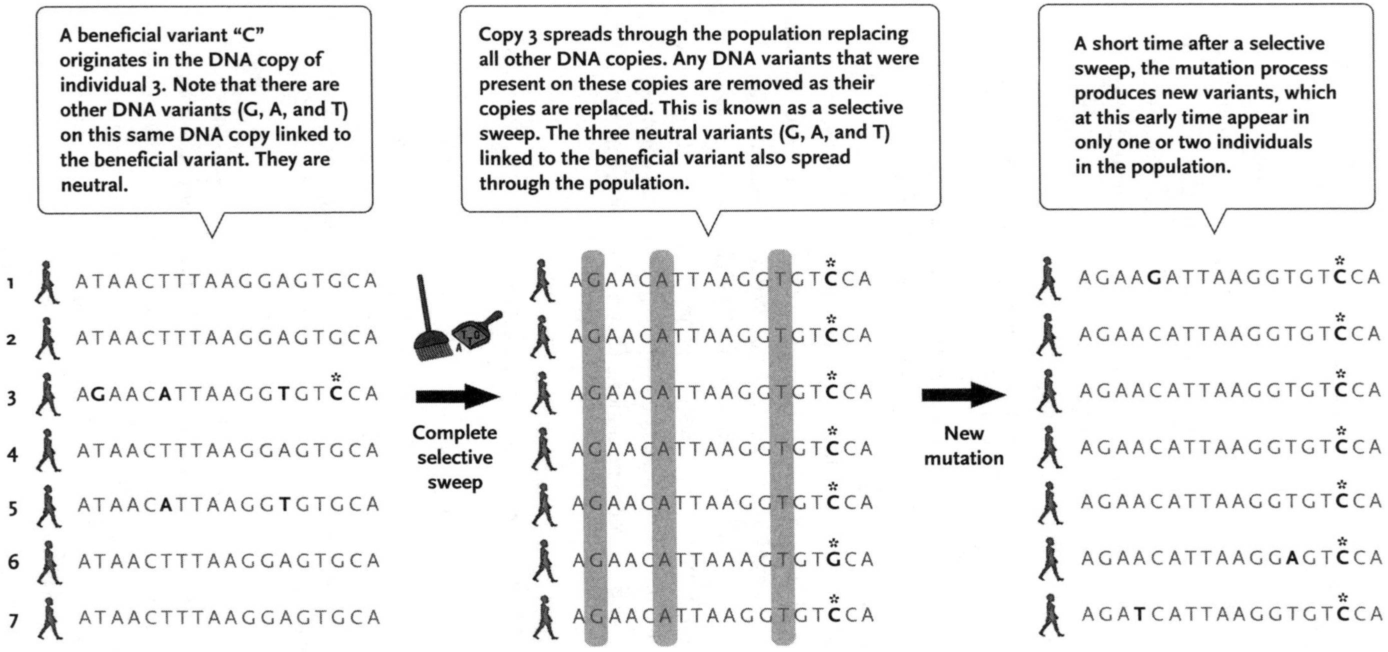

Figure 7.1: A selective sweep.

relatively short time has elapsed since the sweep, these variants tend to occur in very few individuals (e.g., one or two).

We can sum up the effects of a selective sweep by describing the signs left in the DNA sequences of a population at the completion of the sweep. The first sign is that the beneficial variant is now present in most or all of the individuals. Second, the DNA variants that happened to also be present on the sequence bearing the beneficial variant are now present in all sequences in the population, due to hitchhiking. Also, immediately after the sweep, the population has few or no variants near the selected site (due to the "sweep"). The third sign may or may not be present. It depends on how much time has elapsed since the completion of the selective sweep. If not much time has elapsed, then the region around the beneficial variant will be entirely lacking in DNA variants at any site. In contrast, if a longer amount of time has elapsed after the sweep, variants will be present in the population near the beneficial variant but will mostly be present in a single individual (as in Figure 7.1) or only a few individuals.

Collectively the various signs left after a selective sweep are known as a signature of selection. This signature forms the basis for various methods developed by evolutionary geneticists to detect genomic regions underlying population-level adaptations. As in the case of species-wide adaptations, searches for positively selected genes have proceeded in two ways— through a candidate single gene approach and a full genomic scan approach designed to discover gene regions that underlie human adaptations.

SINGLE GENE STUDIES IN ONGOING EVOLUTION: MALARIA

Before whole genome SNP data sets (or sets of DNA variants) became available in the early 2000s, candidate gene studies were the only route to discovering adaptations. These studies focused on a specific gene known to be associated with a specific feature or disease, or a gene suspected to have been under positive selection in the past. The gene was then fully sequenced in those different populations in which the feature or disease was known to be present or absent, and researchers applied methods to detect the signature of selection (usually the signs of a complete selective sweep). Candidate gene studies had a number of notable successes.

One candidate study focused on the gene DARC, which codes for a protein "flag" that sticks out from the surface of red blood cells. For many decades it has been known that while non-Africans usually have this protein, in a large part of the sub-Saharan Africa population the protein "flag" is entirely absent on the red blood cells.[3] This unusually marked difference in the presence or absence of the protein between two geographic regions led researchers to speculate that natural selection may have played a role in creating the difference. It was subsequently found that the protein's presence acts like a beacon for *Plasmodium vivax* parasites to invade red blood cells, causing vivax malaria, a form of malaria different from the more virulent falciparum malaria that afflicts Africans. The absence of the *DARC* protein in Africans was found to provide complete resistance to vivax malaria. Further detailed research led to the detection of a DNA variant (located in the master regulatory region of the gene) that completely shuts off the gene, accounting for the missing protein on the red blood cells of most Africans.[4]

With the advent of DNA sequencing technologies, population geneticist Anna Di Rienzo at the University of Chicago planned a study to detect if the version of the *DARC* gene in Africans had truly been spread through the African population by natural selection.[5,6] The researchers determined the DNA sequence of thousands of bases surrounding and including the DNA site where the resistance variant is located, sequencing this region in a large group of Africans as well as non-African peoples. They discovered the presence of a strong signature of a classic selective sweep in the African DNA sequences but such a signature was completely absent in non-Africans. The DNA sequences in Africans showed reduced variation (due to a "sweep") that extends for several thousands of bases on either side of the beneficial DNA variant. Thus besides the DNA difference between Africans and non-Africans at the site of the beneficial variant, they were also found to be distinct at multiple other DNA sites surrounding the site. This is exactly what would be expected if natural selection had rapidly spread the beneficial DNA variant to all Africans in relatively recent times. In fact, they estimated the onset of the sweep to have occurred slightly over 30,000 years ago.

Despite the fact that DNA technologies and population genetics theory were able to clearly reveal that *DARC* had been under strong natural selection in the past, there is one intriguing and somewhat vexing problem to the *DARC* adaptation. When we consider falciparum malaria in Africa, for which we also know of a DNA variant that offers resistance, we see that the parasite that causes the disease (*Plasmodium falciparum*) is widespread in Africa. In fact, the distribution of the parasite in Africa is almost identical to the DNA variant that provides resistance to it, which makes a lot of sense. This, however, is not the case with the vivax parasite. Oddly, this parasite is not present at all in sub-Saharan Africa! This raises some very tantalizing questions. Was vivax originally present in Africa but later eradicated because most Africans developed resistance? Or, as suggested by Di

Rienzo, was there some other unknown pathogen that caused the selective sweep of the beneficial DNA variant in *DARC*, and it just so happened that the variant also conferred resistance to vivax malaria? At present we just don't know the answer to this *DARC* mystery.

DIGESTING FRESH MILK

Yet another candidate study focused on the variation in the ability to digest the sugar in milk known as lactose.[7] During digestion, lactose is chemically broken down into two simple sugars by the enzyme called lactase. Most people in the world are only able to digest lactose until about the age of five. The enzyme lactase is produced in most people's bodies until this age (roughly corresponding to the age at weaning in traditional cultures) at which point the gene is turned off. However, in some populations, the gene is not turned off and continues to produce lactase throughout adulthood. They are eternal children, when it comes to milk! Anthropologists have determined that these populations include northern Europeans, some Arabians, and several African tribes. Since we know which gene codes for lactase, it was rather straightforward to sequence this gene in multiple populations, choosing both lactose-tolerant and -intolerant populations.

Researchers successfully found a very strong signature of selection in the gene, and this signature was specific to the populations where lactose tolerance was prevalent. In fact, this signature of selection extended for almost 1 million bases on either side of the lactase gene, even encompassing other unrelated genes.[8] The extreme length of the signature indicated that selection on lactase was strong and occurred very rapidly. In fact the studies showed that in Europe, the selective sweep began only about 6,000 years ago. Archeological evidence from domesticated cow fossils, as well as pottery containing milk residues, confirms that cows were being increasingly used for dairying purposes in Europe about this time.[9]

Although the extensive signature of this selection gives us information about the strength of the lactase selective sweep, it poses a big problem for locating the actual beneficial mutation, which becomes something of a needle in a haystack. Nevertheless, painstaking work has revealed that a single mutation in Europeans is almost always present in people with the ability to digest lactose. Surprisingly, the beneficial DNA variant in Europeans was not found within or very close to the lactase gene itself, but 14,000 bases away, within a totally different gene. Additionally, the beneficial variant is in a non-coding portion of the genome, meaning it does not change the structure of the lactase enzyme in any way. Instead, experimental studies have shown that the faraway beneficial variant increases expression of the lactase gene into adulthood. Basically, it keeps the gene switched on.

What then is the underlying beneficial variant that permits lactose tolerance in some Arabian and African tribes? To track down this development in Africa, geneticist Sarah Tishkoff and her students pursued research on lactose tolerance by conducting field research in East Africa from 2000 to 2005.[10] She asked volunteers from diverse tribes to drink glasses of milk and then immediately monitored their blood (through a finger prick) looking for a rise in glucose levels that would indicate lactose tolerance (see Figure 7.2). After lactose-tolerant individuals were identified, their lactose

Figure 7.2: Above is geneticist Sarah Tishkoff as she gives lactose solution to a group of Maasai women in Tanzania to test for a genetic basis of lactose tolerance. Below are Maasai in Tanzania with their cows, which they rely on for meat and milk (Upper photo by Sarah Tishkoff; lower photo by Nova Fisher).

genes (and the genomic region surrounding it) were scrutinized back in the lab using blood samples collected from each such individual. Their collection procedures were approved by an institutional ethics board, research permission from the local governments, and through the signed consent of each individual. They found not just one but several unique DNA variants in the lactose-tolerant Africans. Intriguingly, these different DNA variants are all located within the same approximately 100-base region, known as an enhancer region, where the beneficial variant in Europeans is located. Experimental studies showed that these variants also increase the expression of the lactase gene.

Subsequent studies of Arabians by a group of Finnish researchers showed that Arabs with lactose tolerance also had DNA variants in the same enhancer region, extending expression of the lactase gene into adulthood.[11] Yet again, these variants appeared to have evolved independently of the lactose tolerance variants in Africa or Europe.

There are several important lessons to be learned from the milk studies. First, as with *DARC*, beneficial DNA variants do not always appear within the actual protein-coding portion of a gene, but can appear in non-coding areas that are responsible for regulating the genes (i.e., turning them on and off, or increasing or decreasing a gene's product). Furthermore, regulatory mutations can be quite far from the gene they affect.

Another lesson is that the genetic underpinnings of adaptive features can very well be a result of different beneficial mutations in different geographic populations. If you recall, the independent evolution of the same feature in different populations is called evolutionary convergence. Whenever evolutionary convergence occurs, it shows powerful evidence of the evolutionary process, and specifically natural selection. In evolutionary studies we can rarely carry out laboratory experiments of the type chemists and physicists do. But convergence is like an ancient evolutionary experiment. In the adaptation to milk drinking, the experiment was run in three different populations at slightly different times and the results were the same: the emergence of lactose tolerance. Except there is a twist: each population did it their own way!

LIMITATIONS OF SINGLE-CANDIDATE GENE STUDIES

Despite the important information that can be obtained through single-candidate gene studies, the utmost care must be taken to account for the effects that ancient population dynamics can have on a set of sequences. Any population change in the past, such as reductions or increases in size, geographic expansions, divisions and mergers—and we can be almost certain that some or all of these things happened in our past—can have effects on the pattern of DNA variants in a set of sequences. These effects are the signature of past demographic history. The problem is in teasing apart the signature of past demographic history from the signatures of selection. To give an example, rapid population growth after a bottleneck (period of very low population numbers) will increase the number of variants in a set of sequences that are unique to different individuals. This pattern is almost identical to the pattern that occurs in phase three of a selective sweep, where variants appear in only one or two of the DNA copies in the sample. How then, do we know what a pattern is telling us? The problem is exacerbated by evidence that indicates human populations have been growing rapidly since about 50,000 years ago.

Let's look at another example. There is a gene that enables some people (known as tasters) to distinguish the bitter-tasting chemical present in Brussels sprouts, broccoli, kale, and other cruciferous vegetables. However, people with other versions of the gene (known as non-tasters) cannot taste this bitterness. When populations become partially isolated from one another, or subdivided, this can leave a pattern of DNA variants in a set of DNA sequences that looks very much like the pattern of variants resulting from a form of natural selection called balancing selection. In balancing selection, two or more beneficial DNA variants are spread to large portions of the population since all offer benefits. This form of selection is like a juggler with many brilliant-colored bowling pins in the air at once. Researchers working on one recent candidate study on the bitter-taster gene (dubbed *TAS2R38*) had a particularly tough time teasing the signature of balancing selection from the signature of population subdivision in this gene.[12] It's likely, however, to have been a case of balancing selection. The taster version of the gene is likely adaptive because it allows people to detect potential toxins in plant foods (which can be damaging to the thyroid gland). Detection of toxins must have been important from very early times in our evolution since the taster variant is estimated to date back to over a million years, even antedating the evolutionary transition to our modern human species.[13] One mystery is why some people are non-tasters. But this may be just due to our naiveté. New research suggests that so-called non-tasters, and even people having other known versions of the gene, may actually be able to detect a variety of other potentially toxic chemicals present in other plant foods.[14]

Although we have seen several successes from single-candidate gene studies, we can also see that the interference between a signature of past demographic history and a signature of selection presents a major challenge for revealing the genetic adaptations within populations. Most people agree that the solution to this problem lies in the examination of more genes—actually many more genes! Indeed, examining full genomes is a way to account for the effects of ancient population dynamics.

SCANNING THE GENOME FOR THE SIGNATURE OF SELECTION

Studying genomes in their entirety allows researchers to control for the effects of population dynamics and thereby throw authentic signatures of selection into relief. All the dynamic changes that happened in ancient human populations must have affected all regions of the genome—not one gene, two genes, or three genes. Indeed these changes acted across all individuals in a population and therefore had similar effects across the genome. Thus, there will only be a single signature of ancient population dynamics, even if the population may have had a very complex and dynamic history.

In contrast, the effects of natural selection are localized to specific genes or regions within the genome and will be unique depending on how the natural selection operated. For example, *DARC* has a signature of selection in the African population but not in non-Africans. Lactase, by contrast, has a signature of selection restricted to Europeans and to a few dairying tribes in Africa as well as to some populations of Arabians. The thought is that if it is possible to "subtract" the genome-wide signature from ancient demography, then individual genes and genomic regions underlying human adaptations will stand out.

Since 2005, numerous scans for signatures of selection across full genomes have been carried out, aided by genome-wide data collected by the HAPMAP Project (Haplotype Map of the Human Genome), an international public/private research collaboration, and a private venture known as the Perlegen project led by the Perlegen Sciences company (no longer a functioning company). These projects have screened over three million SNPs in more than two hundred people from several broadly defined world populations including Yoruban, Han Chinese, Japanese from Tokyo, African American and European American groups.

The scans examine millions of SNPs in these genomic data sets section by section, applying methods that measure how the DNA variants are patterned among the different individuals in the data set. For example, one method measures if variants in the set of sequences usually appear in single individuals, a potential sign of a selective sweep. Another method searches for an unusual absence of mutations in and around a specific area, another potential sign of a recent selective sweep. Indeed, researchers usually apply several different methods to measure the different effects left by a selective sweep. This helps to increase the chances of pinpointing a genomic region underlying a population adaptation.

The measures obtained for each section of the entire genome are then used to construct a genome-wide distribution (see Figure 7.3). This illustrates the full spectrum of effects across the human genome as represented in the population set of DNA sequences. According to theory, the set of individual genes within the genome that have undergone selective sweeps will be pulled into the tails of the genome-wide distribution, usually the regions that fall within the highest or lowest 1% of the statistical measures that researchers applied across the genomic data. These genome regions are called outliers, and are believed to be enriched for genes that experienced selective sweeps. The remaining 99% of the distribution is believed to represent signatures left in the sequences by the history of ancient populations.

There have been many successes stemming from the analyses of full genome data sets, with several thousand potential gene regions identified

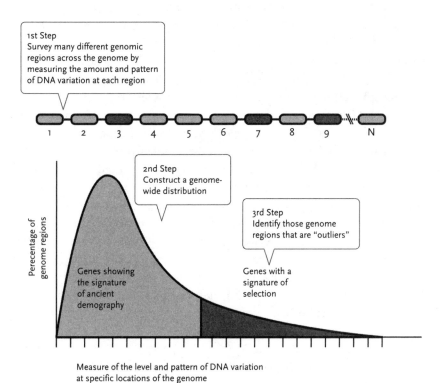

Figure 7.3: A genome scan that uses the levels and pattern of DNA variation across the genome to identify possible regions in the genome shaped by positive natural selection. This figure shows the three steps followed to detect these regions.

as outliers. So what have we learned? One reassuring thing is that many of the old familiar genes from candidate studies, such as *DARC*, still stand out with strong signatures of selection even when embedded in the backdrop of the entire genome, indicating that our methods are capable of pinpointing genes behind human adaptations.

The results from genome scans for population adaptations and the scans for species-wide adaptations (discussed in Chapter 5) also share important similarities. If we put the outlier genes from the population scans into categories, we notice that these overlap considerably with the categories of genes that are also hypothesized to have undergone adaptation during our divergence from chimpanzees. These include genes involved in chemosensory perception, olfaction, resistance to infectious diseases, and spermatogenesis and fertilization genes. This suggests that over the entire course of human evolution there have been persistent environmental pressures for which beneficial variants in these genes were continuously favored. This continuity is especially easy to understand for human genes that provide resistance against infectious diseases, since bacteria, viruses, parasites, and other pathogens are always evolving counter-tactics.

One new category to emerge in population genome scans contains genes involved with cellular metabolic processes. Metabolism is the sum of all chemical reactions in the body. This would include genes involved in modifying and metabolizing proteins, and in metabolizing carbohydrate and phosphates. Recent human evolution has involved migrations of people out of Africa into regions of Europe, Asia, and the Americas, where humans encountered entirely new climactic pressures as well as novel foods. And about 10,000 years ago, in the Fertile Crescent in Southwest Asia, a new agricultural way of life was slowly invented. This practice later spread to groups in other geographic regions. Together, these shifts undoubtedly presented new selective pressures for human populations. In spite of this, pathogens are being found to have played a much more important role in the evolution of our genome than climate or diets. A recent study in 2011 showed that the different pathogens encountered by humans, as we spread to different regions of the world, was the single largest selective force shaping our genes, more so than differences in climate or in foods eaten.[15]

HAIR, SKIN, AND TEETH POP OUT IN GENOME SCANS

Skin color and hair texture are easily recognizable ways in which people in different geographic populations differ. To date, more than twenty genes thought to influence skin pigmentation have been detected in genome scans with strong signatures of selection,[16,17] but it is likely that considerably more genes influence pigmentation. The sheer number of genes found so far is striking; up until quite recently the number was predicted to be only around four.[18] It serves to highlight the possibility that many other human features may be influenced by many genes. Anthropologists once thought that the first colonizers to leave Africa quickly evolved the light skin we see in Asians and Europeans. (The selective pressure for lightening of the skin in northern latitudes is widely thought to allow sufficient amounts of vitamin D to be synthesized by the skin.) But surprisingly, light skin in Asians and Europeans appears to a large extent to have evolved independently after their evolutionary split from each other by the process of evolutionary convergence, that same process that led to lactose tolerance being evolved independently by Europeans, Arabian populations, and dairy-dependent African tribes. The evidence lies in the fact that many pigmentation genes with signatures of selective sweeps in Europeans are not the same pigmentation genes showing signatures of selective sweeps in Asians.[17,19] In 2013, using sets of pigmentation genes identified in genome scans, the evolutionary geneticist Jorge Rocha at the University of Porto (in Portugal) has estimated the initial onset of lightening of the skin to around 30,000 years ago, at least for some shared pigmentation genes between Europeans and Asians. However, for most of the genes that were under selection in Europeans, lightening of the skin appears to have occurred surprisingly recently, between 19,000 and 11,000 years ago, well after the split between Europeans and Asians.[20] It appears that Europeans only achieved their fully lightened skin tones well after their initial expansions deep into Europe.

Genome scans show that the gene *EDAR* has a strong signature of positive selection in people of East Asian descent.[16] The gene is thought to be important in the development of hair follicles. The favored DNA variant in *EDAR*, not found in Europeans or Africans, alters an amino acid in the protein encoded by the gene. To help discover the function of this amino acid change, researchers plucked hairs from people in Japan, Thailand, and Indonesia with and without the *EDAR* variant.[21] When viewed under the microscope, people with the variant had hair strands with thicker cross sections compared to people that lacked the variant, demonstrating a strong association between the common variant found in East Asians and hair follicle thickness. The variant has also been found to be strongly associated with hair straightness in East Asians. To more precisely pinpoint the function of the DNA variant found in East Asians, the geneticist Pardis Sabeti and her co-researchers at Harvard University have inserted the altered gene into mice.[22] Indeed, the mice had thickened hair follicles. However, the story became more complicated since the mice were also found to have smaller mammary fat pads, increased branching of mammary glands, and more numerous sweat glands on the bottoms of their paws. Their research also led them to conclude that the *EDAR* variant arose around 30,000 years ago in Central China and then spread to other East Asians and later to Native Americans. They reasoned that any of the various effects of the East Asian DNA variant could have been important in spreading the variant. (The EDAR DNA variant is also associated with unique tooth features, like shovel-shaped incisors, large crowned teeth and extra cusp on molars, features common in Asian populations.[23]) Despite these many effects, researchers hypothesize that its effects on outward appearance—for example, on hair texture and amount of mammary fat—may have figured prominently in the features people looked for in potential mates. Perhaps, when choosing their mates, thick and straight hair or reduced breast size was seen as a particularly alluring trait.

THE GENOMIC HIGHWAY

The researchers riding the genomic highway do not obey speed limits. Already they have reached the limits imposed by the HAPMAP and Perlegen databases.[24] These databases, developed in the mid- to late 2000s, have surveyed the genetic variation in our species only very broadly in three or four world populations. Because of this, they cannot identify the adaptations of any population not included, of which there are many. Moreover, the databases are crude because they do not contain a record of every mutation actually present in each individual in the database. This is because the SNPs surveyed in the project were selected merely because they were known, perhaps through the initial human genome project in 2001, or through a small survey of SNPs in Europeans for disease studies. The databases therefore give us a biased understanding of the extent and nature of genetic variation in the world's populations. (As described in the previous chapter, this bias is called ascertainment bias.) Second, the databases do not provide a full record of the SNPs for each individual screened. This is simply because the projects did not determine the full sequence of A, C, G, and Ts in the genome for these individuals, only screening for well-known variants. This results in uneven screening of genetic variants across the genome; some regions have dense numbers of SNPs and other regions have far fewer. In fact, low density of SNPs in these studies is partly to blame for why some genome-scanning studies were unable to identify the selective sweep in the *DARC* gene related to vivax malaria.[25]

In response, researchers interested in studying the unique adaptations of specific populations have extended the SNP data of the HAPMAP and Perlegen databases to new populations. For example, recent studies of the genes underlying the ability to live under low-oxygen conditions at high altitudes have surveyed SNPs in both Andean (Bolivia and Chile) and Tibetan plateau populations that have adapted to life at altitudes well above 2,500 meters (8,000 feet), and, for some Tibetans, at twice this altitude.[26,27] For each of these populations, researchers compared the high-altitude genome-wide SNP profile with the genome SNP profile of nearby lowland dwellers. Results indicate that Andeans and Tibetans adapted to life at high altitudes independently from each other, but the list of genes identified as having undergone positive selection are associated with such physiological functions as sensing of oxygen levels by cells, nitric oxide synthesis (important for dilating blood vessels), sensing cellular energy, and hemoglobin production.[26] These candidate genes will serve as points of departure for more detailed investigation to see what their exact function is in each population. The sets of genes that showed signatures of positive selection in each high-altitude population appear to be mostly different, a finding not totally unexpected since the two populations have different physiological adaptations.[28] High-altitude Andeans generally have higher numbers of red blood cells and higher hemoglobin counts, making their blood viscous. (Hemoglobin is the molecule that carries oxygen in the blood; having more of it enables the blood to carry oxygen more efficiently at low atmospheric concentrations.) High-altitude Tibetans, however, have lower hemoglobin counts, equivalent to people living at sea level, but seem to have other adaptations like more rapid resting ventilation. However, oxygen levels in their blood appear to be low relative to Andeans. It appears Tibetans, therefore, must have other mechanisms to deal with low oxygen levels in their blood: perhaps the rate of blood through the lungs is faster, perhaps blood is delivered to the body's tissues faster, and perhaps the movement of oxygen molecules into the tissues is more rapid. There is some evidence that Tibetans have such adaptations.[28] Although Andeans have increased red blood cell counts and hemoglobin levels in their blood, a question arises as to how they avoid the possible negative health consequences such as abnormal blood clotting, stroke, and poor maternal blood flow to the fetus. Having located some of the genes that appear to underlie adaptations to life at high altitudes should let us begin to dissect exactly how populations have adapted in different ways to such a life.

Yet another high-altitude population has recently been investigated using a genome SNP-screening method—the Amhara highlanders in Central Ethiopia who live up to 3,505 meters (11,500 feet) above sea level.[29] Although their physiological adaptations to life at high altitudes are not known, the genomic scan found several genes across their genomes that appear to have undergone positive selection but did not undergo selection in lowlanders and that appear related to tolerating hypoxic (low-oxygen) conditions. Interestingly, most but not all of the genes were different from those identified in Tibetans and Andeans, and so it seems that adaptation to life in the Amhara Plateau of Ethiopia also occurred by mostly separate genetic routes.

In the most significant development in this area, advances in technology have enabled a new project known as the 1000 Genomes Project, launched in 2008, which aims to solve many of the limitations of the HAPMAP and Perlegen studies. The project originally aimed to carry out full sequencing—the complete genome sequence of A, C, G and Ts—for one thousand people from diverse world populations. Already the mission has changed to over two thousand individuals (see what I mean by speeding?). In addition, the project hopes to discover differences among human genomes, such as duplications or deletions of entire genes. The benefit of the project is that it will determine all genetic variation contained in these populations at a frequency of 1%, meaning that if a genetic variant is present in at least one person out of one hundred in a particular population, the 1000 Genomes Project will find it. Since it detects all genetic variants across a genome, it is not biased in its survey of genetic variants. Already, in 2010, the pilot phase of the project completed the genomes of 179 individuals. The project detected fifteen million gene variants, more than half of which had never been detected before.

Initial results from the 1000 Genomes Project have caused us to reconsider the process of adaptation in human evolution.[30] Results indicate that signatures of selective sweeps, of the type we have been describing thus far, are relatively rare in the human genome. To be precise, we are talking about the type of signature left after a beneficial DNA variant arises in a population and is then rapidly spread to all or most people in the population. Adaptation in some genes like *DARC* and others may have followed the classic selective sweep process, but in general human adaptation may have proceeded through a somewhat different process—indeed, for some time evolutionary geneticists have been noticing that selective sweeps seem on the whole to be rare.

Another hint of this problem came from surveys of SNP variation in more extensive populations than were surveyed in the HAPMAP and Perlegen databases. We expected to find strong genetic differentiation between different geographic populations, especially in the genomic regions underlying biological functions important for survival in those regions. Contrary to expectations, and despite the detection of some gene regions showing strong differences, there appears to be a general absence of strong genomic differentiation between populations from different geographic regions. Most of the genetic differences between geographic groups stem from the time of their ancient migrations out of Africa into Europe, Asia, and the Americas, as well as the differences resulting from restricted exchange of mates correlated with the distances separating groups.[31]

How to explain these unexpected findings, especially when we know human populations do show physiological and anatomical differences that appear to be adaptations to different environments? How to make sense of a dearth of genomic signatures of selective sweeps that might underlie these adaptations? The simple answer appears to be we have been looking for the wrong signature of positive selection.

So is there a more subtle road for us to drive, one that stretches beyond the selective sweep? Researchers traveling this road, in fact, have uncovered a process of adaptation called a *soft sweep*. There are several elements to a soft selective sweep.[32] First, it has long been hypothesized that most human features are not simply based upon a single gene (such as the case with the lactase enzyme or the protein hemoglobin) but are polygenic, or influenced by many genes. This has been confirmed by recent genome-wide studies, which indicate numerous genes underlie features such as height, body mass index, basal metabolic rate, blood pressure, and lipid and glucose levels in the blood. Furthermore, it is believed that each separate gene has relatively small effects on the overall feature. This means that a functional mutation at any single gene would not need to sweep through the entire population for the population to adapt to its environment. Indeed, experimental studies have shown that all that seems to be needed is a partial sweep—a soft sweep—of the functional variant, provided that the same sweep occurs simultaneously at all or most genes that influence the feature. This is known as *polygenetic adaptation*.

Another necessary ingredient for a soft sweep is that functional variants (DNA changes) that influence the feature must already exist in the population at each of the genes. This is different from the classic selective sweep, or *hard sweep* as we can now call it, in which a beneficial variant appears for the first time at a single gene in a single individual and then spreads to everyone else in the population. In the polygenetic adaptation model, the functional variants are already present in many individuals even at the start of the soft sweep. In fact they can be present in as few as 10% all the way up to 50% of the individuals in the population. To take an example of a feature we can easily relate to, height, we might examine variation in height for a large group of people (for the sake of example we can use data from the Quaker population in Boston). We would see only a few people that would be very short (let's say 156 centimeters, or

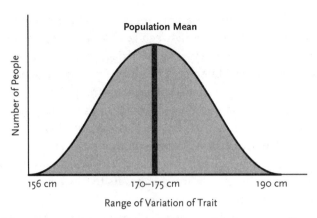

Figure 7.4: Continuous variation in height, with few people being very small or very tall but with most people tending to the average height.

about 5′1″) and similarly only a few people that are very tall (let's say 190 centimeters, or almost 6′3″). If we measured enough people, we would end up with a continuous distribution like that seen in the illustration (Figure 7.4). We would see that more and more people will have heights that approach the average height, with the most people very near or at the average height of about 170 and 175 cm (5′7″ to 5′9″). This is known as continuous variation. Indeed, many polygenic features in a population will vary in a continuous fashion.

Recent studies estimate that there are likely to be hundreds of DNA sites spread over numerous genes that influence height. Each DNA site alone may influence height by a rather miniscule amount, perhaps only a few millimeters. For sake of explanation, let's suppose that spread across the genome there are a total of five hundred DNA sites that influence height and each affects height by only two millimeters.[33] At each site there can exist either a DNA variant that favors an increase in height or a different variant that favors a decrease in height (black and gray in Figure 7.5). You can see that for each site, the tall and short variants are found in different proportions in the initial population prior to selection (upper graph). For site 1, the tall variant is found in 60% of the people but the short variant is found in 40%. If we look at site 2, the tall variant is found in only 30% of the people and the short variant in 70%. Looking over all the five hundred DNA sites, we will find the tall and short bases are found in varying percentages of the population at each particular site.

Now let's suppose the environment of this population changes in a way that favors an increase in height. Natural selection will thus favor individuals who have inherited sets of DNA variants that increase their height. To see how the height of a population can change by polygenetic adaptation,

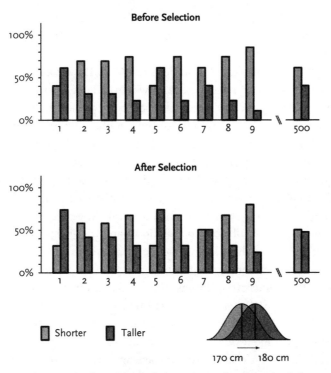

Figure 7.5: A soft sweep for the polygenetic trait associated with height shifts a population toward a 10-centimeter greater average height.

let's look at the population after some generations have passed (lower graph in Figure 7.5). Now, you can see that because tallness is favored in the population, natural selection has caused each DNA variant for tallness to become 10% more common in the population, while those for shorter stature have become less common (by 10%) in the population. So, at site 1 the tall variant is now found at 70% of people and the short variant is now in 30% of people. Such a shift occurs at site 2 and all 500 sites.

To see how this results in a shift in average height from an initial average height of 170 mm (5′7″), imagine the improbable scenario of a person inheriting the tall variant at all five hundred sites. If this happened it would cause an increase in their height of a whopping 1,000 mm (or 100 cm) and making that person 8′8″ tall! This will not happen of course. The more realistic scenario is that a person inherits tall variants at only a subset of the total five hundred sites, increasing their height by a more modest amount, depending upon how many of the five hundred sites have tall variants. Let's now assume that natural selection spreads the tallness variants to only 10% of the population. Under this scenario, the average height of the population will increase by 10% of 100 cm (10 cm), making the new average height 180 cm (5′9″). You can see this in the bottom of Figure 7.5, where the continuous distribution for height in the population has shifted to the right by 10 cm.

Because polygenetic adaptation does not leave the strong signature left by a selective sweep at a single gene, new methods of detection have been devised, with some success. One new method is known as environmental correlation. The method assumes that human phenotypes are significantly correlated with aspects of geography or aspects of the environment. For example, skin pigmentation and body mass are significantly correlated with the incidence of UV light radiation and latitude, respectively. Therefore, the SNPs underlying these features should similarly be correlated. In 2008, numerous SNPs within pigmentation and metabolic genes were surveyed in a large diverse sample of world populations, and these SNPs were found to show subtle shifts in their frequencies correlated with specific and varying aspects of climate.[34] For pigmentation genes, the shift in the frequencies of certain SNPs depends on the intensity of UV radiation at different latitudes.

For numerous SNPs within various metabolic genes, shifts in the frequencies of certain SNPs depend on the particular latitudinal position of a population, most likely due to differences in heat or cold stress experienced at these different latitudes.[34] In contrast, in control tests, SNPs that we know are unrelated to metabolism or pigmentation show no correlation with varying environmental factors. Interestingly, many of the SNPs in metabolism genes are known to be associated with diseases such as type 2 diabetes and obesity. Perhaps the biological processes that allow tolerance to climatic changes (or extremes) can also influence one's susceptibility to certain metabolic diseases.

Susceptibility to salt-sensitive hypertension also seems to be correlated with latitude, or more specifically temperature. A hypothesis raised in 1973, called the sodium-retention hypothesis, suggested that the higher prevalence of hypertension in African Americans compared to European American descendants was due to their adaptation to hot equatorial climates by conserving water loss and thereby maintaining the body's core temperature.[35] Consistent with this hypothesis, SNPs located in the *AGT* (angiotensinogen) and *CYP3A5* genes, both known to regulate water loss in the kidneys, were found in a 2004 study to be strongly correlated with distance from the equator, becoming more common by roughly 10% for every ten-degree shift closer to the equator.[36] For both metabolic genes, it appears that soft adaptive sweeps with only subtle shifts in the frequency of SNPs within these genes have occurred to allow adaptive adjustments to climatic variables. Future research will likely increase the number of SNPs detected with frequency shifts correlated with climate variables.

The life ways of people around the world vary from region to region with respect to the particular type of eco-region they live in (e.g., dry, humid, polar), the type of subsistence strategies they practice (e.g., hunter-gathering, agriculture, pastoralism, horticulture), and the main components of their diets, such as roots and tubers, meat, fats and milk, or cereals. A study in 2010 led by Anna Di Rienzo[37] looked for subtle shifts in the frequency of SNPs in different world populations to find out whether the shifts were correlated with any particular type of eco-region, subsistence practice, or dietary practice.

The results indicated that certain SNPs are correlated with the dietary specializations of different world populations. For example, it was noticed that certain SNPs showed increased commonness in populations that rely on tubers and roots (e.g., yam, cassava, taro, cocoyam, and potato), which are high in carbohydrates. Interestingly, the SNPs correlated with such a diet are located within genes that help in the digestion of carbohydrates like starches and sucrose. Other SNPs found to be more common in populations reliant on root and tubers compared to populations with non-tuber/root diets included SNPs located in a gene known as *MTRR*, which functions in folate (vitamin B9) synthesis in the body. This seems related to the fact that tubers and roots are very low in folate (with deficiencies of this vitamin known to cause birth defects), and so individuals in these populations may have evolved enhanced ways to synthesize their own folate. There is another interesting correlation too, this time with a diet focused on cereal grains. There is an increased commonness of a specific SNP in the *PLRP2* gene (a gene which codes for a lipid-digesting enzyme) in populations that specialize in eating cereal grains, but this SNP is less common in populations in which cereal grains are not the main dietary component. The correlation is found even when populations with these contrasting diets are adjacent to each other (see Figure 7.6).[37] The SNP at increased commonness in cereal grain-eating populations is thought to improve the enzyme's efficiency in breaking down the main lipid found in cereal grains and was likely promoted by natural selection in these populations.

A somewhat different way a genome may evolve to allow people to adapt to different diets is through the duplication of genes, especially if the gene codes for an enzyme important for the digestion of a specific and important food. The anthropologist George Perry at the University of Texas found that the amylase gene (*AMY1*), which produces the starch-digesting enzyme in saliva, has duplicated itself to varying degrees among different populations.[38] People with greater numbers of copies produce more amylase in their saliva compared to people who have less copies of the gene. (There is a slightly different form of amylase we produce in our pancreas,

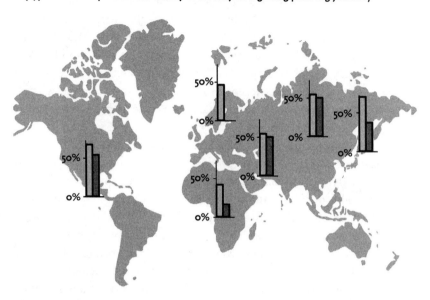

Frequency of SNP in the PLRP2 gene
(hypothesized to promote the activity of the enzyme digesting plant triglycerides)

▨ Populations specializing in cereal grains

▨ Populations not specializing in cereal grains

Figure 7.6: This map shows that an SNP in the *PLRP2* gene thought to help to break down the triglycerides in plants is found to be more common in populations with diets focused on cereal grains (light gray) than in populations with diets not specializing on cereal grains (dark gray). (Redrawn from Hancock, et al. 2010, p. 8928.[36])

but a different gene codes for this amylase.) Interestingly, when they looked across different world populations, they found that the number of copies of the *AMY1* gene in a person's genome is associated with the amounts of starch traditionally eaten by people in that population. So, people in populations that adopted diets of starchy agricultural foods (within the past 10,000 years or so) have more copies of the salivary amylase gene than folks from societies that traditionally were hunter-gatherers. This difference exists even for populations that are relatively closely related, indicating that natural selection must have been a relatively powerful force in driving the dietary adaptation. As an example, while the Yakut of northeastern Siberia—who consume meat and dairy products and obtain their food through fishing, hunting, and a nomadic herding way of life—possess on average about five copies of the *AMY1* gene, the Japanese, who adopted agricultural ways early on, have on average seven copies of the gene in their genomes. What is more striking is that this pattern is repeated in diverse populations around the world, in which populations that have adopted diets high in starch typically have more copies of the *AMY1* gene than neighboring populations that rely on low-starch diets.

RIDING THE GENOMIC HIGHWAY INTO THE FUTURE

As we've seen, genome-wide studies can have advantages over single gene studies in helping to identify outlier gene regions that may relate to human adaptations. However, merely being an outlier does not necessarily mean that a gene region underlies a human adaptation. It only tells us that the region has a pattern of DNA variation that is peculiar in some respect compared to the pattern of variation in the rest of the genome. In fact it is quite probable that many outliers will be unassociated with human adaptations. On the other hand, some of the genome regions that fall within the genome-wide distribution (see Figure 7.3) may in fact be found to underlie human adaptations. For example, the soft sweeps of SNPs correlated with aspects of climate, diet, or subsistence activity, identified in environmental correlation studies, do not leave signatures that would stand out as outliers in a standard genome-wide scan for genes under selection. Yet many of these SNPs do likely underlie important human adaptations.

Nevertheless, the searches for hard and soft sweeps ultimately suffer from the same problem. That is, for many of the genomic regions identified as possibly adaptive (which number in the thousands), we have little to no idea about their function or their effect at the phenotypic level. But we should remember that natural selection acts on the phenotypic expression of a gene or SNP, because this is what increases or reduces an individual's ability to survive or reproduce. This problem has led Joshua Akey, at the University of Washington, to describe genome-wide studies as a "hatchet" that identifies pools of candidate genes, while advocating for the detailed "scalpel work" to be carried out to verify which of these many genes represent actual adaptations.[39] Contrary to the advice of the Gryphon in *Alice in Wonderland*, who impatiently urges "No, no! The adventures first, explanations take such a dreadful time!" it is imperative that we do the difficult "scalpel work" to first verify adaptation at candidate genes, and then to more fully comprehend their adaptive benefits, before we start telling "just so" stories.

How do we apply the scalpel? In-depth study of the hundreds of candidate genomic regions emerging from studies is certainly a place to start. To strengthen the case that a region underlies an adaptation, it is necessary to identify the specific population or populations in which the hypothesized adaptation occurred, and determine exactly when in human evolution the adaptation occurred. Finally, it is most important to determine the actual functional effect of the genomic region under question. These follow-up studies very likely will require studying the candidate region in additional populations. For example, detection of selection for lactose tolerance in small dairying tribes in East Africa would not have been identified without studying genetic variation at the lactase gene specifically in these populations.

For genes like lactase, *DARC*, and others, there is information from functional studies. For example, the SNP in the lactase gene among lactose-tolerant individuals was confirmed in laboratory experiments to up-regulate the production of the enzyme lactase whereas the SNP in lactose-intolerant people does not. Another example comes from the skin pigmentation gene *SLC24A5*. Its phenotypic effect was determined by studies on zebrafish because genetic variation in the gene underlies differences among fish in the number, size, and density of melanosomes (skin cells that produce the dark pigment melanin) and ultimately their outward coloration.[40] After screening for *SLC24A5* in the HAPMAP populations, researchers discovered that Europeans have a signature of a classic sweep at this gene.[41] The further identification of an amino acid alteration in Europeans (but not in Africans) appears to verify that it contributes to their lighter skin. This gene represents a good example of the general need to identify phenotypic effects of candidate genes. Without the functional information from our experimentally friendly Zebrafish (common in pet stores), *SLC24A5* would have remained one of many genes of unknown significance floating in the "outlier" pool.

Some genes that underlie human population adaptations have certainly undergone the classic selective sweep type of adaptation as described in the beginning of this chapter. But it is becoming clear that polygenetic adaptation, involving soft selective sweeps at many different genes, each of them having a small effect on a complex feature, has been perhaps more important than single gene adaptations. Over the next few years, we will begin to get a better understanding of exactly how widespread polygenetic adaptation has been in allowing human populations to adapt to different environmental conditions. This should help us gain a better understanding of the evolution of complex features among populations such as height; diseases such as hypertension, diabetes, obesity; and skull shape and skeletal form. Indeed, learning more about the genetic adaptation of complex features that show considerable differences among living populations will help us more thoroughly understand the genetic basis of our species-wide adaptations in such features as brain size and structure, cognition, language, bipedalism, hand and finger proportions, and manual dexterity.

CHAPTER 8

Kissing Cousins—Clues in Ancient Genomes

In Gorj County, Romania, in 1952, a group of bones, including an almost, complete skull of a woman, was found in the multichambered Cave of the Old Woman (Peştera Muierii). Along with the bones was found an advanced kit of tools made from stone and bone. In 1998, a skeleton of a young boy, about age four, was discovered in the Old Mill Rock Shelter (Lagar Velho) in Portugal. Only fragments of the skull remained, but the skeleton itself was better preserved. Evidence suggested a burial. The body was painted with red ocher, was presumed to be wrapped in cloth, and had a pierced shell of a periwinkle snail placed below his neck as an ornament. Enclosing the grave was a ring of vertically placed animal bones and large stones. Both fossils proved to be the remains of very ancient peoples that lived in Europe, dating to 30,000 and 24,000 years ago, respectively, providing researchers with the opportunity to make some provocative new claims about ancient humans.

Both individuals possessed a curious mix of anatomical features. In 1999, the Portuguese archeological team led by João Zilhão at the Univeristy of Barcelona described the Old Mill boy's skull remains and skeleton as possessing a bony ear region with anatomical features intermediate between those of Neandertals and humans, and having a distinctly robust and stout skeleton like Neandertals.[1] In 2006, the Romanian team, led by American anthropologist Erik Trinkaus at Washington University, described the old woman's skull as having many modern features—delicate jaws, smaller front teeth, more delicate nasal openings into the skull—but also with a handful of Neandertal features, most notably a well-formed "bun-like" protuberance on the back of the head known as an occipital bun.[2] Based on these observations, the researchers concluded that the old Romanian woman and the young Portuguese boy were offspring of matings between Neandertals and anatomically modern peoples—human hybrids, if you will.

ANCIENT GENOMES IN ANCIENT FOSSILS

For over 150 years, two different beliefs about Neandertals had persisted in some form or another: either Neandertals were our relatives, or they were not. The interpretation of Neandertals as an odd offshoot from the human lineage was largely fueled by the ideas of the influential French paleontologist Marcellin Boule who, in the early 20th century, described them as dumb-witted, lumbering brutes with stooped postures and imperfect gaits (Figure 8.1). That idea persisted, perhaps unconsciously, through many analyses that have highlighted the "peculiar" features unique to Neandertals—the occipital bun, the strong protruding brow ridges, the forwardly jutting face—contributing to the belief of Neandertals as an altogether separate side branch from the lineage that ultimately led to humans today.

The idea that Neandertals categorically split from the human lineage hundreds of thousands of years ago has persisted. Even today, common representations of the human evolutionary tree (Figure 8.2) still show a sharply arched and diverging branch that leads to Neandertals and away from the human lineage, virtually screaming "We've got nothing to do with each other. Once we split ... we split!"

Figure 8.1: On the left, Marcellin Boule (1861–1942), French paleontologist. On the right, is a diorama displayed at the Field Museum in Chicago during the 1930s of a Neandertal family in Gibraltar inspired by Boule's interpretations of Neandertals as being brutish, dim-witted, bent-kneed, and hunched over.

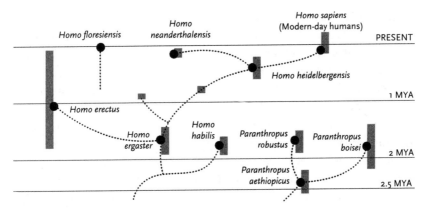

Figure 8.2: Close-up view of a human evolutionary tree.

To say that claims of hybridization between Neandertals and anatomically modern humans raised eyebrows is an understatement. In the past, some anthropologists saw a place for Neandertals as possible ancestors of modern humans, but as we've seen, there is an overwhelming body of genetic and fossil evidence that anatomically modern humans originated in Africa, and subsequently spread out of Africa. It was commonly held that modern humans then replaced any archaic humans who had migrated out of Africa at earlier times. This would have included replacement of the Neandertals, who inhabited Europe and western Asia from around 200,000 years ago to as recently as 30,000 years ago. This description of modern human origins, known as the "out of Africa" model (though several variants of this model have been proposed), suggests that Neandertals are our close evolutionary cousins who, despite sharing a common ancestor with us, split from our lineage several hundred thousand years ago and then went extinct when modern humans—humans that anatomically look like you and me—spread out from Africa and replaced them with either no interbreeding or only very minor amounts of interbreeding.[3]

Another camp has argued for a Multiregional Continuity model of modern human origins, in which the Neandertals are seen as one of many archaic human populations that were spread out over the Old World—namely Africa, Asia, and Europe.[4] These populations are proposed to have been connected by gene flow through interbreeding at shared contiguous borders. According to this model, there were no sharp splits of separation between populations where they persisted in isolation for long periods of time. If there were anatomical differences between groups, these were based on adaptation to local conditions—for example, Neandertals with their short, stocky, and muscular bodies and wide nasal apertures and large noses were well suited to life in Ice Age Europe—gene flow and natural selection served to tie the regional populations together so that the groups would all evolve into modern form.

The exuberant debate over whether Neandertals were replaced or inter-bred with modern humans has endured for decades in the anthropologi-cal field. More recently, a middle ground hypothesis has emerged, the *mostly* out of Africa hypothesis.[5] Its proponents suggest that after mod-ern humans evolved to their modern form in Africa, they *mostly* replaced Neandertals and other archaic humans outside Africa. Thus, when modern humans spread out of Africa about 50,000 years ago, they sometimes inter-bred with Neandertals and other archaic humans when they met.

The skeletons of the young boy from Portugal and old woman from Romania, with their mix of anatomical features, stoked the flames of the debate, thrusting it into the forefront of our minds. If the skeletons proved to be actual hybrids, it would provide some answers to the tantalizing ques-tions about the fate of the Neandertals. But it is tremendously difficult, if not impossible, to say with any degree of certainty whether a fossil comes from a hybrid solely on the basis of skeletal features. After all, we have little idea what such a hybrid would look like in its skeletal characteristics, and as we shall see, it is often not possible to recognize known hybrids among living animals from their bones alone. Resting solely on the fossil evi-dence, the question over possible interbreeding between modern humans and Neandertals remained a mystery. But the recent extraction of ancient genomes from our archaic cousins has provided a new method to determine, definitely, when humans and Neandertals diverged in our evolutionary past, when and how they dispersed, and whether they in fact did interbreed.

Using paleogenomics, researchers are starting to unravel tantalizing questions that we have grappled with for decades. What was the nature of our evolutionary links to our archaic ancestors? To what extent did we interbreed with our archaic cousins and where did these possible matings occur? Were there evolutionary advantages to mating with our archaic cousins? To what degree are our presumed unique human traits, such as advanced language and cognition, truly unique to us? Are the fossils of the young boy from the Old Mill from Portugal and the old woman of the cave from Romania really direct proof of our first reproductive encounters with our ancient cousins?

In 2009, the plot thickened with the help of paleogenetics, when fossil evidence of another new and distinct type of archaic human was discov-ered, which had lived in Asia during the same time as the Neandertals. This second long-lost archaic cousin is known as a Denisovan after the cave site in the Altai Mountains of southern Siberia (see Figure 8.3), where only a

Figure 8.3: The Denisova Cave in the Altai Mountains in southern Siberia. Ancient DNA extracted from a fossil finger bone discovered at the site revealed that a previously unknown hominin group lived in Asia, distinct from the ancestors of modern humans and distinct from Neandertals (Photo: Bence Viola, Max Planck Institute).

bony tip of a pinky finger and two molars were found. The bony fingertip comes from a geological layer dated to around 50,000 years ago, and has no telltale anatomical signs. It was therefore a mystery as to whether it belonged to an early ancestor of modern humans or some other archaic human relative, perhaps a Neandertal. Not long after its discovery, Svante Pääbo and his team at the Max Planck Institute in Germany scooped out most of the interior of the bone, extracted its DNA, and in 2010 established a draft of its full nuclear genome.[6] Unexpectedly, the genome from this finger bone showed a genetic distinctness that excluded it from a place on that lineage leading to modern humans and also excluded it from a position on the lineage leading to Neandertals. Instead, the genome from this anonymous finger fossil led to the unexpected revelation that modern humans had a second archaic cousin, aside from the Neandertals, that had lived outside Africa. DNA extracted from one of the two molars discovered in the cave substantiated this finding, and because it derives from a second and older individual, indicated that the cave was inhabited by a group of these distinct ancient hominins. (Interestingly, only the molar, which is unusually very large and robust compared with the molars of both Neandertals and humans, offers clues as to the possible anatomical robustness of the Denisovans.) As with Neandertals, researchers quickly wondered if anatomically modern humans ever interbred with Denisovans.

In one of the most significant advancements in the study of anthropology, we can now employ a new and powerful "telescope," helping us look backwards to peer into the genomes of our ancestors. The telescope is actually the new field of paleogenomics, consisting of methods for extracting, reconstructing, and then analyzing full genomes from fossilized bones. Fossils form by a gradual process called mineral replacement in which the original biological materials are replaced by minerals such as silica or calcite. However, only the oldest of fossils are actually entirely replaced by minerals, and some more recent fossils, less than 100,000 years old, still contain small amounts of biological material from which researchers can recover DNA. The field of paleogenomics allows us to look backward in time, to see genomes in our ancestors, and to obtain definitive answers about their evolutionary ancestry. With the tools of paleogenomics, anthropologists can even discover hitherto unknown ancestors without stepping outside the laboratory.

A BRIEF LOOK AT THE HISTORY AND CHALLENGES OF ANCIENT DNA RESEARCH

Advancements in DNA techniques over the past twenty years have helped revolutionize our understanding of human evolution by allowing the extraction and analysis of ancient DNA from hominin fossils. Among the myriad challenges in working with DNA tens of thousands of years old are two paramount primary hurdles. If DNA is present at all in the fossil, it is so significantly degraded that it exists only in small DNA fragments (on the order of 50 to 150 bases) so that in order to reconstruct a respectable stretch of the DNA, or an entire genome (billions of bases long), the short fragments need to be painstakingly joined together. Second, ancient DNA in fossils is significantly contaminated by extraneous DNA that had infiltrated the fossils over thousands of years—DNA from bacteria, other animals, and most perniciously, the DNA of the human handlers of the fossils. Fortunately, scientists have forged ahead in developing sophisticated laboratory techniques for extracting ancient DNA and minimizing the effects of contamination, as well as computer-based methods that are able to piece together small remaining bits of the DNA into longer DNA segments that can be compared with the DNA of living individuals.

The first advancement in ancient DNA research came in 1997 when Pääbo and his ancient DNA team published a short DNA region, totaling about four hundred bases in length, of the mitochondrial genome of a Neandertal.[7] The ancient DNA had been extracted from the Neandertal-type fossil, discovered by quarrymen in 1856 in a cave in the Neander Valley in Düsseldorf, Germany. By the mid-2000s, methods for working with ancient DNA had advanced to the point of determining entire mitochondrial genomes of Neandertals, approximately 16,500 bases in length. To date, there exist numerous mitochondrial DNA short DNA sequences from Neandertals, and also more than seven complete mitochondrial genomes, allowing us insight into how our ancient cousins varied from region to region. In an even more remarkable feat, in 2006 large segments of the nuclear genome of Neandertals were first obtained.[8,9] Yet our feelings of triumph were tempered shortly after when analyses showed that DNA from living humans had crept in, contaminating some of the data.[10]

Persistence in mitigating the many challenges of working with ancient DNA paid off and soon led to bigger breakthroughs. In 2010, Pääbo and his team published drafts of the entire genomes of two extinct hominins—the entire nuclear genome of a Neandertal[11] and the entire nuclear and mitochondrial genomes of a Denisovan.[6] The Neandertal genome was derived from three fossil bones from a cave in Vindija, Croatia. These ancient genomes gave us our first comprehensive look into the DNA blueprints of our extinct relatives, giving us genetic evidence of when we split from our extinct cousins and tantalizing clues about when and where we would meet again. Yet even these genomes were not of the quality and accuracy of genomes determined for living people.

One important way we can ensure the accuracy of a genome—so that we know confidently what the order of A, C, G, and Ts is within that genome—is to determine its sequence of bases over and over again in independent experiments. This process generates what we call "genome coverage," and the first drafts of ancient genomes were sequenced to only less than two times average coverage. Increasing coverage would prove difficult because it would require greater quantities of DNA extracted from the fossil, and authentic ancient DNA in fossils only exists in minuscule amounts. In fact, it's been found that the fraction of genuine ancient hominin DNA in hominin fossils is most often less than 1%, and rarely exceeds this.[12]

While Pääbo's team has been at the forefront of generating ancient hominin genomes, they have also been pioneers at developing new laboratory techniques to improve the accuracy of ancient genomes. Whereas the draft genomes were produced through a protocol relying on double-stranded DNA, a process found to be inefficient in using the precious miniscule amount of starting DNA, one of their more recent techniques started building the ancient genomes by first unzipping the double helix of the ancient DNA into single strands, thereby doubling the amount of starting ancient DNA molecules.[12] The method proved to be efficient in other ways too, and ultimately was found to increase the recovery of ancient DNA by ten times, a great leap from such small beginnings.

Using such techniques, in 2012 Pääbo's group published a new version of the Denisovan genome from the same individual, but this time of much higher quality.[12] Then in 2014, the group published a high-quality genome of a second Neandertal individual.[13] This time the ancient DNA derived from a toe bone of a Neandertal woman found in the very same cave in southern Siberia from which the Denisovan harkened and so is geographically very distant from where the draft Neandertal genome is from (Vindija, Croatia). The Neandertal toe was found in a somewhat lower geological stratum then where the Denisovan finger was found and so is believed to be slightly older. These ancient genomes are now determined to thirty times genome coverage, guaranteeing that each base has been verified an average of thirty times. As such, the quality of these genomes far surpasses that of the initial drafts of 2010 and rivals the quality of genomes obtained for living people, where the original source DNA abounds and contamination issues are much less severe. In fact, the accuracy of these ancient genomes is so high that we can tell when parts of its genome were inherited from different parents, although we can't tell exactly which parent passed them on. It is also remarkable that the errors in the sequence due to contamination are down to less than 1%. Also in 2014, Pääbo's group determined the genome of a third Neandertal, albeit at much lower coverage: an infant from Mezmaiskaya Cave in the Caucasus.[13] This genome, along with the Croatian and southern Siberian genomes, could now be compared to learn important information about genetic diversity in Neandertals.

As an aside, it's remarkable to note how freely available these ancient genomes have been to other researchers. Unlike in the old (and not so old) days of paleoanthropology, when some scientists jealously guarded their fossil discoveries, drawing the resentment of researchers and obstructing progress, researchers in genomics usually make genomes available to all scientists through public databases, and sometimes even release the raw data before they themselves have been able to fully scrutinize and publish their data. As an example, Pääbo's team released their second Neandertal genome in March 2013 for analysis well before they penned their own publication on the secrets within this ancient genome.

GOING OUR SEPARATE WAYS

When did we split from our ancient cousins and begin to live separately from them? These questions have been studied for decades by comparing differences in the anatomy of fossil Neandertals and fossils of early modern humans with little to no resolution. But what do ancient genomes extracted from fossils say?

To gauge the time when Neandertals split from the human lineage, Pääbo's research group reasoned that Neandertals split from the human lineage much more recently than the splitting of chimpanzees from humans: some five or more million years ago.[11] The first step was to locate all the DNA sites at which humans had a derived base, say an A base, whereas the chimpanzee had a different base, say a T base. (The chimpanzee base was assumed to be the ancestral base, in which case the DNA change from T to A must have evolved on the lineage leading to humans.) They then looked at all the same DNA sites (over two million sites in all) in the Neandertal genome and determined the proportion of sites at which Neandertals also had the human base. They found that this was true for 18% of the total sites, meaning Neandertals had an 18% special DNA closeness to humans measured in reference to a genome from an individual from the Yoruban population in West Africa. The same test was then applied to genomes of different human individuals from around the world and they found (as you'd expect) that these humans had greater degrees of evolutionary closeness to the Yoruban human reference—29.8% (Han Chinese), 29.7% (French), 29.3% (Papuan), and 26.3% (San Bushman from Africa). As you can see, the Neandertal genome is more distant from the Yoruban individual—8% to 12% so—compared to any of the diverse human individuals. After some calculations, and assuming a chimpanzee divergence of either 8.3 or 5.6 million years ago based on two different calibration points for when orangutans split off (thus yielding maximum and minimum time estimates), Pääbo's group estimated the population split between Neandertals and modern humans to have occurred between 440,000 and 270,000 years ago. These dates are similar to the splitting times of 383,000 to 275,000 years they found based on the higher-quality 2014 Neandertal genome from southern Siberia. Besides providing estimates based on the phylogenetic mutation rate, they also used the "twice-as-slow" *de novo* mutation rate (described in Chapter 8), which yielded estimates twice as old. However, as described previously, researchers are currently unsure as to which of these mutation rates is more accurate. Resolving this problem in the near future will be essential since the time-scale of events in human evolution is so vastly different according to the two different rates.

Even if we base our discussion on dates derived using the phylogenetic rate, the genomic estimates for the split between Neandertals and modern humans is still very old, especially when compared to traditional inferences based on fossils that place the earliest Neandertals at around 200,000 years ago.[14] These relatively old dates for when Neandertals and modern human populations parted ways indicate that Neandertals indeed represented a distinct population separate from modern humans, and also that we had been evolving in isolation from them for a considerable amount of time. The genetic distinctness of Neandertals holds up when partial Neandertal genomes are examined from fossils from the El Sidrón site in Asturia, Spain, to the Vindija site in Croatia (Central Asia) to Mezmaikaya Cave, north of the Black Sea in the Caucasus. This means that Neandertals, even over a large range, seem to have belonged to a rather tightly knit genetic population and that modern humans and Neandertals represented distinct populations that were reproductively isolated from each other for much of their evolutionary history.[15] Interestingly, Neandertals have been estimated in the studies by Pääbo's team to have had a very low *effective population size* (as you recall, a measure of how their population behaves in a evolutionary sense) of roughly 2,500, or roughly one-third the *effective population size* of modern humans. Such a small effective population size means that the random evolutionary forces of genetic drift must have been a very significant force not only shaping their genome, but also influencing their morphology.[15]

The Denisovan genome shares newly evolved DNA bases with Neandertals that we do not see in the genomes of humans, indicating that Denisovans and Neandertals were still very much part of the same archaic population after they split from our lineage as long ago as 400,000 years ago—although, over a period of roughly 150,000 years, Denisovans and Neandertals became increasingly isolated from each other and finally parted ways with one another about 250,000 years ago.[6,12,13]

Remarkably, at the Denisova Cave there are also tools and other artifacts indicating ancient modern human occupation in addition to the already mentioned Neandertal fossils from the cave, although no definitive fossils of modern humans are known.[16] This indicates that three different hominin forms—modern humans, Denisovans, and Neandertals—lived at the cave site in southern Siberia, though probably at different times. During the time of occupation, from about 50,000 to 30,000 years ago, the local regions surrounding the cave appear to have consisted of pine and fir forests, though at times expanding into more meadow-like areas. The tool types show continuous development, with lower levels at the cave containing typically Neandertal-like tools grading into more sophisticated and varied tools usually associated with modern humans. Additionally, personal decorations consisting of beads made from tubular bone, pendants made from mammoth tusks, and ornaments made from mollusk shells are also found, which indicate a spiritual and symbolic social life at the cave.

In one of the upper layers of occupation, Russian archeologists uncovered the pieces of a shimmering bracelet made from dark green chloritolite stone, a stone transported from some distance. At present, we are not certain of the history of occupation at the Denisova Cave, and which of the three different hominins were responsible for the different artifacts found there. However, with the genomes of two distinct hominins from the site already determined, the Denisova Cave might represent a unique "case-study" for us to explore the associations between each of these hominin groups and the technological and possible symbolic and spiritual life each group led, and to determine when each resided at the cave and if they ever directly crossed paths there.

Until relatively recently, it was thought that the Neandertal lineage emerged about 200,000 years ago. But under a new scenario illuminated by its ancient genome, we now know that the Neandertals of Europe and Asia appear to have had a considerably older history, extending as far back as possibly 400,000 years ago or more. Furthermore, fossils suggest their features evolved by a drawn-out process, known as the accretion model, in which the specialized Neandertal features were added by a gradual step-by-step process over time.[17] But who might be the first Neandertal ancestors in Europe?

In 1997, a remarkable site was discovered in Atapuerca, Spain, called the Pit of Bones (Sima de los Huesos), where at the very bottom of a forty-eight-foot (fifteen meter) vertical shaft and another forty-foot (twelve meter) slope, and over a quarter-mile from the current entrance to the pit, were found the largest collections of hominin fossils anywhere in the world. These comprised approximately 6,500 bones and teeth belonging to at least twenty-eight separate individuals dating to 400,000 years ago or older.[18,19] Interestingly, these archaic skeletons show some incipient Neandertal features, and although the age of this site is under some debate,[20] the ancestry of Neandertals might be found in these Atapuercan ancient hominins. With this possibility in mind, Pääbo's team extracted DNA from a fossil thigh bone from the site, ultimately reconstructing its entire mitochondrial genome in 2013. This is remarkable as the fossils are hundreds of thousands of years older than any previous hominin fossil from which ancient DNA has been extracted. The unexpected findings could throw a wrench into the interpretation that the Pit of Bones people represent direct ancestors of Neandertals, since their mitochondrial genome is more closely similar to the mtDNA genome of Denisovans than to that of Neandertals.[21] At this point, it remains a mystery as to what explains the unexpected mitochondrial link between the ancient peoples from southwestern Europe and the far-away ancient Denisovans of southern Siberia. If nuclear genomic evidence becomes available for the Atapuercan fossils as well as from other comparably old fossils in Europe and Asia, we should get a clearer picture of the locations of the earliest roots of the Neandertals as well as Denisovans.

ABBA-BABA—A NEW MODEL OF HUMAN ORIGINS

Through the 2000s, the strict replacement model maintained a strong hold on anthropology research, despite models of modern human evolution that actively sought to include the possibility that modern humans might have interbred with archaics. But even before the first publications of the genomes of archaic human ancestors, evidence was mounting that indeed there may be at least a small degree of archaic DNA in our genomes. By the mid-2000s, a small group of population geneticists and their research teams, including Michael Hammer at University of Arizona, Jody Hey at Temple University, and Damian Labuda at the University of Montreal had independently discovered regions in the human genome that had gene lineages that were proportionally very ancient when compared to the majority of our genome. These regions were also in chromosomal regions that had not experienced the usual amount of reshuffling that had affected other genomic regions.[22-25]

Since, over the course of many generations, the chromosomes that we carry usually get broken up and reshuffled with other bits of chromosomes from other individuals by the process of crossing-over, as mentioned earlier, there is a direct relationship between evolutionary time and how much reshuffling takes place; the longer the evolutionary time, the more reshuffling. All regions of a chromosome will usually show roughly the same level of reshuffling. However, in the 2000s small chromosomal regions were being discovered that were devoid of the "usual" reshuffling. After rejecting some alternative explanations for this, scientists hypothesized that these regions did not show reshuffling because they had been isolated from the modern human gene pool for extended periods of time. The regions then reentered the gene pool at a later time through interbreeding of modern humans with archaic populations. Unfortunately, while population geneticists were fairly accepting that these results indicated that some interbreeding had taken place between modern humans and the archaic populations like Neandertals, anthropologists steeped in the strict replacement model were not so open. They either felt that such regions were somehow atypical and not reliable, or felt that some form of natural selection (perhaps balancing selection, which can sometimes maintain genetic variants within populations for very long periods of evolutionary time) could better explain these deep lineages. Some of the genes predicted to have been passed to us from Neandertals—like the brain-related gene *microcephalin* or the neurologically functioning *MATP* gene—proved eventually to be absent in the Neandertal genome, leaving their origins unexplained at present.[11,26] Yet, some predictions were right. Seven years prior to the determination of the Neandertal draft genome in 2010, Damian Labuda's research group discovered a surprisingly ancient variant of the dystrophin gene (a gene that codes for a muscle protein) in about 9% of Europeans and Asians. This ancient gene has now been verified as present within the Neandertal genome and likely came from them.[27]

When the full genomes of the archaic Neandertals and the Denisovans became available in the early 2010s, the hypothesis that modern humans interbred with archaic populations could finally be rigorously tested. If we assume that interbreeding between modern humans and archaics took place in the past, and that some human populations interbred with Neandertals to a greater extent than other populations, these populations should share more DNA similarities with Neandertals. One test of interbreeding and consequential intermixing of archaic and modern genomic material is dubbed the "ABBA-BABA" test and compares the genomes of two modern humans with the genomes of the archaic hominin (in this case the Neandertal) as well as the chimpanzee (see Figure 8.4). In the test, researchers examine only those DNA sites in the genome in which individuals from different human populations (let's say an African and a non-African) are known to carry different bases: one carries a base denoted "A," and the other carries an alternative base denoted "B." "A" stands for ancestral, meaning the base is found in the chimpanzee genome and that when we see it in one of the human populations we can assume it was inherited from our last common ancestor shared with chimpanzees. On the other hand, "B" stands for a derived base (not found in the chimpanzee) that is found in the genomes of at least one of the human populations, but not both, and also in the archaic genome. In Figure 8.4, these are represented by site 11 in the ABBA tree (upper tree) and site 7 in the BABA tree. If the Neandertals did not interbreed with modern humans, then it is expected that each human branch will share "B" sites equally with the archaics or that there will be equal amounts of ABBA and BABA sites. On the other hand, if there was interbreeding between archaics and at least one of the two human populations represented in the study, then one or the other human branch will share more "B" sites with the archaics and ABBA-BABA sites will not be equal.

What does the ABBA-BABA test say about our encounters with archaic hominins? With respect to Neandertals, for the first time direct evidence

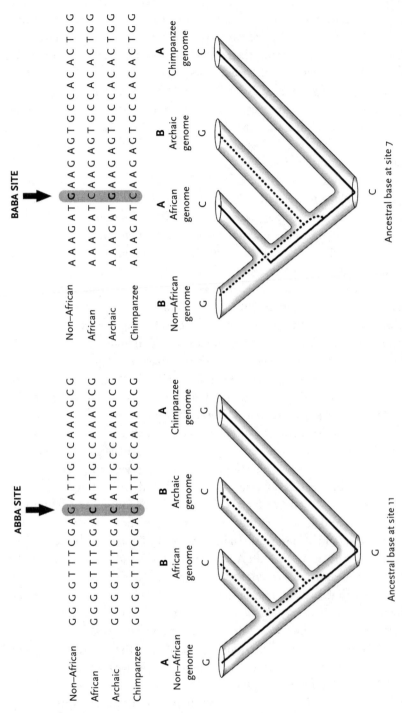

Figure 8.4: The ABBA-BABA test allows researchers to determine if archaic hominins, like the Neandertals, interbred with modern humans and if so specifically which modern human populations were involved.

from an ancient genome tells us that some of our modern human ancestors indeed interbred with Neandertals. Furthermore, the ABBA-BABA test points toward interbreeding between Neandertals and non-Africans (Eurasians) but not between Africans and Neandertals. This is because more derived "B" sites are shared between Neandertals and Eurasians than between Neandertals with Africans. The test can tell us roughly the proportion of Neandertal genomic ancestry in Eurasians. Based on the draft genome generated in 2010, this estimate was that up to 4% of the modern Eurasian human genome is composed of Neandertal DNA, and it appears that both Asians and Europeans share Neandertal DNA equally.[11] This estimate was subsequently revised downward to 2% based on the high coverage 2014 Neandertal genome from the toe bone found at the Denisova Cave in southern Siberia.[13]

What could explain such a pattern—no Neandertal DNA in Africans but very nearly equal amounts of Neandertal DNA in Europeans and Asians? The simplest explanation would be a scenario in which anatomically modern humans encountered Neandertals when modern humans migrated out of Africa, but before they differentiated into Asians and Europeans. The route out of Africa is not certain, but one likely path is through the Sinai region just north of the Gulf of Suez (see Figure 8.5), then into the eastern Mediterranean region and finally into Europe and Asia. At paleoanthropological sites located in the eastern Mediterranean region both Neandertal remains (Shanidar Cave, Iraq, and the Amud, Kebara, and Tabun sites in Israel) and modern human fossils (Qafzeh and Skhul caves in Israel) have been recovered. So did the two groups ever meet in the eastern Mediterranean, and if so, when?

A direct meeting between moderns and Neandertals in the eastern Mediterranean is not entirely clearcut. However, anthropologist Chris Stringer at the Natural History Museum, London, and geologist Rainer Grün at the Australian National University have established possible ages for the Neandertal fossils from the nearby Tabun Cave in Israel using absolute dating techniques (i.e., radiometric analyses) that would seem to indicate that modern humans and Neandertals had overlapped in the area. A nearly complete female Neandertal skeleton from Tabun was found to be slightly older than 120,000 years and a series of seven teeth (also from Tabun), likely from the upper jaw of a Neandertal, were found to be about 90,000 years old.[28,29] The three burials of modern humans from the Skhul Cave fall into a 100,000- to 135,000-year interval, while the Qafzeh modern human fossils have been variously dated from 90,000 years ago to as old as 120,000 years ago.[30] If these dates are correct, Neandertals might have been living "right around the corner" from modern humans, since

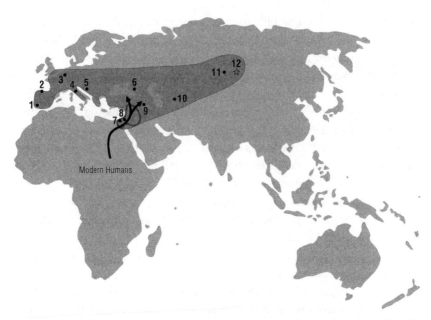

Figure 8.5: Areas of Neandertal geographic distribution are shown in light gray, while the region in the eastern Mediterranean where interbreeding between anatomically modern humans and Neandertals likely took place is shown in darker gray. Solid dots indicate Neandertal sites and also indicate the two early modern humans sites of Qafzeh and Skhul Caves in Israel (dots 7 and 8). The star indicates the Denisovan Cave where skeletal remains were found from which the Denisovan genome was determined. Arrows represent a possible route of modern humans out of Africa into Eurasia. 1. Forbes' Quarry, Gibraltar; 2. El Sidrón, Spain; 3. Feldhofer, Germany; 4. Saccopastere, Italy; 5. Vindija, Croatia; 6. Mezmaiskaya, Caucasus; 7. Skhul Caves, Israel (modern human), and Tabun Cave (Neandertal); 8. Qafzeh Cave, Israel; 9. Shanidar, Iran; 10. Tashik Tesh, Uzbekistan; 11. Okladnikov Cave; Siberia; 12. Denisova Cave, Siberia.

Tabun and Skhul are neighboring caves on the southwest flank of Mount Carmel in Israel and the cave at Qafzeh is not far away (Figure 8.6).

In 2012, a genomic analysis estimated the timing of Neandertal interbreeding with modern humans to fall between 65,000 and 47,000 years ago.[31] This is more recent than the dates for the Skhul, Qafzeh, and Tabun sites as described. However, some researchers have proposed that there were actually two distinct pulses of modern humans out of Africa into the eastern Mediterranean region, an early pulse represented by the Skhul and Qafzeh people around 90,000 to 135,000 years ago (which subsequently receded back into Africa around 75,000 years ago), and then a second pulse some tens of thousands years later around 45,000 years ago.[32] The genomic time estimates for interbreeding would be consistent with interbreeding taking place during the second pulse of modern humans into the eastern Mediterranean. Archaeologist Paul Mellars from the University of Cambridge has described a sharp and sudden change in technology in

Figure 8.6: Mount Carmel in Israel, showing the Tabun Cave, where Neandertal fossils have been found, and the Skhul Cave, where modern human fossils have been recovered. (Photo Claudio Vita-Finzi.)

the eastern Mediterranean during the second pulse, with stone blades, burins used for engraving objects, end scrapers, and perforated shell ornaments proliferating.[33] However, while the results of genome comparisons between humans and Neandertals gives us proof that moderns and Neandertals interbred, and we believe it to have occurred somewhere in the eastern Mediterranean region, the exact time when Neandertals and modern humans first met and interbred and over how long a duration interbreeding took place is still unknown. Many of these questions could be resolved if ancient DNA is recoverable from fossils of ancient humans and Neandertals from the cave sites from Mount Carmel or from the Qafzeh cave, as will very likely become possible in the near future.

MIGRATING, MEETING, AND MATING WITH DENISOVANS

Unlike the Neandertals, the Denisovans are known virtually only by their genome and from scant fossil remains from the Altai Mountains in Siberia. What do tests like ABBA-BABA tell us about modern humans interbreeding with Denisovans? Initial comparisons of modern human genomes with the Denisovan genome in 2010 yielded the astounding finding that modern-day populations living in Australia (aboriginal), and Melanesia, including Papua New Guinea as well as the Philippines, showed the most evidence of interbreeding with ancient Denisovans. Indeed, from 3% to 6% of the genomes of South Asian Islanders are of Denisovan origin.[6,12] So while these people also have Neandertal components to their genomes, the Denisovan fraction is as much as threefold greater. Moreover, no evidence for Denisovan DNA has been found in European populations so far, but more recent studies have detected Denisovan DNA in the genomes of Han and Dai Chinese as well as in South American Indians, albeit at twenty-five times lower levels.[12,13] Denisovan DNA in South American Indians is not surprising since genetic evidence points to a mainland Asian ancestry for indigenous American peoples.

The reason this finding is so astonishing is that Denisovans were discovered from bones in a cave in southern Siberia. You may wonder what DNA from Denisovans is doing within the genomes of southeast Asian tropical peoples. In fact, we do not yet know. One idea is that archaic Denisovan populations were spread over a vast territory, perhaps covering a good portion of eastern and southern Asia—eastern Russia, Mongolia, China, and into South Asia. Then, as the ancestors of humans presently living in the southeast Asian islands migrated out from Africa, and then moved across the mainland of southern Asia, they met archaic Denisovan populations already living there and interbred with them to a limited extent. These modern people then continued on to the islands of southeast Asia carrying some Denisovan DNA with them.

HOW MUCH KISSING BETWEEN COUSINS?

As we compare the ancient Neandertal and Denisovan genomes with more and more modern genomes, we are learning that the locations and times of our hybridization with these archaic hominins were complex. With further research, it is likely that this ancient soap opera will turn out to be even more involuted and knotty. As modern humans migrated they likely encountered archaic hominins in many regions and sometimes interbred with them. When they did interbreed, the archaic bits that mixed into the genomes of modern humans were carried within their genomes like archaic "suitcases" of DNA. The presence and absence of this archaic DNA in modern human populations gives us clues to the routes that modern humans traveled as they left their homeland in Africa, and may be able to tell us how many different dispersals from Africa humans may have made. For example, the nearly equal amounts of Neandertal DNA in European and Asian genomes points to the probability that there was at least one major migration out of Africa and that at least one interbreeding event occurred before Europeans split from Asians. As noted, one possible yet uncertain location for this tête-à-tête is in the eastern Mediterranean. The number of meeting and mating places will most certainly become more complex as the Neandertal genome is compared with genomes from additional human individuals from diverse populations. For example, initial Neandertal genome analyses compared this archaic genome with only eight other modern human genomes. More recently, the Neandertal genome has been compared with almost seventy genomes from Europe, Asia, Southeast Asia, and Africa.[34] Much larger studies are now in progress in which the Neandertal genome is being compared with the numerous genomes being generated by the 1000 Genomes Project. What will these larger analyses reveal?

Although the eastern Mediterranean seems one likely place that Neandertals and modern humans met and interbred, was it the only place? At present, we just don't know. Paul Mellars believes modern humans started colonizing disparate areas in Europe, including regions in present-day Romania, France, Bulgaria, and the Czech Republic, as well as in the Near East in Lebanon and Israel, at relatively early times—in some places as early as 40,000–50,000 years ago.[35] Such an early colonization would have given Neandertals and modern humans numerous opportunities to meet and for them to potentially interact. Mellars and other archeologists see evidence in the tools and other artifacts at some Neandertal sites that could indicate that Neandertals borrowed techniques or styles from nearby modern humans. In particular, the Châtelperronian stoneworking industry at Neandertal sites in northern Spain and western and central France, as well as the Uluzzian industry in south and central Italy dating to between 40,000 and 30,000 years ago, have a mixture of typical Neandertal tools and more sophisticated tools usually attributed to modern humans. It has often been argued that these industries represent cultural exchange between Neandertals and their new modern human neighbors, who had recently expanded into the areas and were now living nearby.[33,36]

Some archeologists propose that Neandertals survived until quite recently—until 30,000 years ago or even more recently—in southern Iberia, in a kind of warm southern European refuge zone.[37] However, much pivots on the accuracy of the radiocarbon dates. In recent research—in which new pretreatment methods are applied that reduce contamination by present-day carbon—the recent dates proposed for the southern Iberian Neandertals are found to be far too recent, and may be as much as 10,000 years older.[38] Considerably older dates, if accurate, would suggest that Neandertals did not linger on in southern Iberia and would reduce the time that modern humans and Neandertals could have overlapped in the region.

Recent investigations have looked more closely at regions in Italy that might have been Neandertal-modern human meeting places. Indeed, in 2012 the anthropologist Silvana Condemi at the Aix-Marseille University in France suggested that Neandertals and modern humans were essentially neighbors for a period of time at geographically adjacent sites in the foothills of the Alps in northern Italy—Neandertals at the Mezzena Shelter and modern humans at the Grotta di Fumane. Such evidence makes it possible that the two interacted and even possibly mated with each other in this area.[39]

But up until now, genomes do not show any evidence that, as modern humans spread into Europe, they mated with Neandertals. Nevertheless, there are some tantalizing questions here. Since our samples of genomes from Europeans from different regions is far from comprehensive at present, it remains possible there are some European peoples that contain larger amounts of Neandertal DNA than others. Or perhaps Europeans and Neandertals did in fact interbreed (in the time range of 30,000 to 50,000 years) in various regions where they met in Europe, but their hybrid descendants have not survived to the present day. For instance, recent studies of ancient DNA from skeletons of early European peoples who practiced hunting-gathering lifestyles found that these people were replaced around 7,000 years ago by farming peoples coming from the Near East.[40–42] The relatively rapid spread of these farming peoples from the Near East into Europe could have largely obliterated any genetic sign of interbreeding between Neandertals and the earliest modern humans to colonize Europe. In the future, we should be able to scrutinize this further by analyzing the full genomes of the skeletons of the early Europeans, antedating the expansion of farmers into Europe, to examine whether and where in Europe modern humans and Neandertals could have met and mated.

We also don't know the extent of Neandertal DNA in the genomes of continental East Asians. The first comparisons showed that Europeans and East Asians had equal amounts of Neandertal DNA.[11] But with increased sampling of peoples in East Asia, researchers have found that some East Asian populations may contain greater fractions of Neandertal DNA than Europeans,[34] which has been confirmed by analyses of the 2014 high-coverage Neandertal genome from the Denisova Cave.[13] It may have been that East Asians had greater contact with Neandertals than did Europeans, which is surprising, since the geographic range of Neandertals has usually been thought to cover Europe into Western Asia. Yet, one scenario could be that the descendants of Neandertal-modern human interbreeding in the eastern Mediterranean could have later migrated into East Asia. In any case, it seems that East Asians must have intermixed with Neandertals after they separated from Europeans. Once we have a better idea of the relative amounts of Neandertal and Denisovan DNA in peoples from different geographic regions—the number or sizes of their archaic DNA "suitcases"—we will have greater power to trace the likely quite complex migration routes of modern peoples out of Africa.

The recovery of the genome from a 20th-century Australian Aborigine, extracted from a hair sample, has also shown that Denisovan DNA exists at a very similar level (around 6%) in these original Australian islanders as in people from Oceania and Melanesia (Papua New Guinea and Bougainville Island).[43, 44] Interestingly, the pattern of Denisovan DNA in Southeast Asian populations is patchy. Some Southeast Asian populations show evidence of interbreeding with Denisovans, while others do not. For example, populations from western parts of Indonesia do not have Denisovan DNA. Thus, the hypothesis that Southeast Asians mated with Denisovans on the mainland of Asia before these modern peoples colonized the islands of Southeast Asia does not seem to fit the evidence. Rather, the intermixing seems to have occurred within Southeast Asia itself, meaning that the Denisovans could have had a range that extended from Siberia all the way into Southeast Asia, yet with an uneven distribution among the islands of Southeast Asia.[44] If further studies confirm the presence of Denisovan DNA in East Asians (for example, Chinese),[12, 13] it could suggest multiple episodes of interbreeding between Asians and Denisovans over a very broad swath of territory in Asia.

At this point in our research, the gene flow seems to have been mostly in one direction. Archaic DNA from Neandertals was passed along into modern human populations, more so than from modern humans into Neandertals.[11] However, this finding could be modified in the future once genomes from more Neandertal fossils are sequenced. It is important to point out that the fractions of Neandertal DNA in modern human genomes are very low. Indeed, the small fractions point to interbreeding rates that must have been quite low too, suggesting the two hominins had a strong avoidance to mating with each other. It could also mean that their hybrid offspring had low fitness.[45] This makes us wonder about the contexts of the reproductive encounters between humans and Neandertals. Were they mutual or forced interactions? Did they involve broader social ties or none at all? How did hybrid offspring fare in terms of their health, and which of the two hominin groups cared for them?

ANCIENT HYBRID ZONES: EVOLUTION'S LABORATORIES

Regions where two different species meet and interbreed are called *hybrid zones*. In the past, many scientists thought hybrid zones were rare or static areas if they existed at all. And many thought they were the result of an abnormal breakdown of the reproductive isolating mechanisms that keep species separate. Hybrid zones are now understood to be common at the geographic borders where closely related species meet. These zones are also dynamic places where two species can pass functionally important genetic material back and forth.

Clifford Jolly has studied hybrid zones for more than thirty years among distinct types of baboons. They are so distinctive that they are easily distinguishable from one another; in fact, there are five types of baboons that form a ring-like series encircling the sub-Saharan African regions. Some researchers have designated each of the distinct types of baboon as separate species, but Jolly has found that zones of hybridization exist wherever the types meet. One might very well argue that these different forms of baboons are not in fact distinct species, because they easily reproduce and produce fertile offspring.[46] Rather, the baboon types might be seen as different geographic "races" or what some researchers call different baboon subspecies. Importantly, the hybrid offspring at hybrid zones will often mate with members of either of the two "pure" parental populations and produce fertile offspring, a process known as backcrossing. DNA sequence evidence collected in the late 1990s by Jolly and his students indicates that these different types of baboons probably began diversifying as long ago as 1.7 million years ago.

There are similar hybrid zones in Mexico, where the distributions of two different howler monkey species meet. In this case, the monkeys are believed to have split from one another as long as three million years ago. Therefore, from a monkey's perspective, it is not surprising that archaic humans like Neandertals and Denisovans interbred with modern humans and produced fertile offspring—especially since their evolutionary split was less than one-third of the time since the split between hybridizing howler monkeys or baboons. Indeed, Jolly has long argued that the African baboons, with their tendency to hybridize wherever their boundaries meet, serve as an illuminating analogy for the reproductive dynamics between emerging humans and our extinct archaic cousins.[46]

Hybrid zones have been described as evolutionary laboratories, since new genetic combinations are constantly being formed in any new hybrid offspring. And these new gene combinations that come together are parsed constantly by natural selection. Some combinations of genes shuffled together in the hybrid offspring might be detrimental and quickly lost through the hybrid's death or inability to mate. But some offspring might have their genes shuffled in such a way that they become beneficial, and these individuals will be reproductively successful and pass the beneficial gene combinations on to future generations. Since hybrid offspring often backcross or mate with a member of one of the "parental" species, the new gene combinations borne by the hybrid can spread throughout the parental species. This is called gene flow across the hybrid zone, and it behaves like a factory generating new genetic raw material for evolution.

With respect to modern humans and archaic hominins, hybridization probably occurred at the "wave of advance" as the moderns with their more sophisticated technologies colonized new environments. Such hybridization need not have been merely some hanky-panky at the frontier, but could easily have been an area of evolutionary experimentation, where the colonizing populations gained certain beneficial genetic variants from archaic hominins. After all, when modern humans spread out of Africa, they were largely adapted to tropical environmental conditions, foods, and pathogens. Since their spread appears to have been relatively fast in evolutionary terms—a timespan of thousands of years—there would have been limited time to adapt to new conditions out of Africa. Contact with new pathogens, foods, and environmental conditions in Europe and Asia would have proven challenging to the colonizing modern humans. Hybrid zones, where modern humans could have obtained genes from Neandertals or Denisovans who already had been adapted for life outside of Africa for several hundreds of thousand years, would have been extremely significant. Archaic genes could have been shuffled into the modern human genome, and if beneficial, spread by natural selection to more and more modern humans.

Is there any evidence for beneficial genes coming into the modern human genome through hybridization with archaic hominins? What are the functions of these genomic regions and did they offer important functional benefits to modern humans? Our answers to these questions are limited, since most of the current analyses have not been designed to pinpoint the genomic locations of archaic genes, and therefore it is hard to say much about the functional consequences of archaic genomic regions. Most analyses of the Neandertal and Denisovan genomes have been statistical measures that estimate the percentage of archaic hominin DNA in the modern human genome, and do not determine the locations of the archaic bits. Also, the archaic genomes have been compared with relatively few complete modern genomes, so it has not yet been possible to map how widespread archaic bits are in the human genome. These limitations will soon be less evident, however, as more and more genomes of geographically diverse humans become available to compare with archaic genomes and analyses begin to dissect the function of the archaic bits of our genome.

Some researchers have already hypothesized that some archaic bits in our genomes may have important functions. Immune genes are highest on the list. Mike Hammer and his group have detected several different immune genes in modern humans that are almost identical to their archaic counterparts and may have originally come from archaic humans. These immune regions also show much less shuffling compared to other human genes, meaning they are recent genetic imports into the genome. One of the genes is the *OAS1* innate immune gene that has an ancient variant in Oceanians and people from eastern Indonesia, dating to almost three million years old. It is identical to the region in the Denisovan genome.[47] Hybridization must have occurred between Denisovans and the ancestors of Melanesians and Indonesians in Southeast Asia, since the *OAS1* variant is not known outside this region.

A second immune gene in humans, *STAT2*, shows a variant almost identical to the *STAT2* gene in the Neandertal genome. While the *STAT2* variant is found throughout Eurasia, it is not very common, being present in only about 5% of people there. In Melanesia, however, the *STAT2* gene variant seems to have had a beneficial function, though we are not sure what it was. It has become ten times more common in Melanesians than in Eurasians, indicating natural selection favored it in Melanesia.[48]

Though the exact functional importance of these archaic genes in our species' genome are not yet known, they could very well have played important roles in helping modern humans fight local pathogens in regions outside Africa. Several of the well-known human leukocyte antigen (HLA) genes located on chromosome 6, which help attract T white blood cells and natural killer cells to destroy pathogens in the body, have also been hypothesized to have come from either Denisovans and Neandertals.[49] These genes do have known functions, although it's unclear at present whether they offered a distinct advantage to modern humans. Proposed HLA genes from Denisovans are most commonly found in populations in Asia, rather than in Europe, while HLA genes from Neandertals are common in populations from both Europe and Asia. This makes sense if Asians interbred with Denisovans only after they had already separated from Europeans, and if the ancestors of Eurasians interbred with Neandertals before they split into separate populations (as has been surmised).

One possible confounding aspect is the role that balancing selection is known to play on some immune genes. Balancing selection is the form of natural selection that maintains different gene variants at nearly equal proportions in a population over very long periods of evolutionary time. For example, some HLA gene variants have been maintained, or "balanced," since the time of their evolutionary origin over ten million years ago in the common ancestor of humans, chimpanzees, and gorillas. Thus while some of these immune genes are likely to have been introduced into the human genome by hybridization with our archaic cousins, others may simply be ancient immune genes that evolved deep in the past and hold fast because they are important to us. Teasing apart these alternative explanations remains central to truly identifying genes "imported" from Neandertals or Denisovans.

DO WE HAVE FOSSIL HYBRIDS?

So were the boy from the Old Mill in Portugal and the old woman from the cave in Romania really hybrids? It is not possible to tell for certain. It might be unlikely, however. There is a low probability of discovering a hybrid as a fossil, when fossils are in the first place very rare. Hybridization between modern humans and our archaic cousins itself must have been quite infrequent since the percentage of Neandertal DNA in European and Asian populations ranges from only 1% to 2%, and the percentage of Denisovan DNA in Asians ranges between 3% and 6%. Studying hybrid zones between different monkey species and other animals alive today might serve to give us further insight into the issue of how likely it is to uncover actual hybrid individuals in the fossil record. Hybrid zones are very restricted in space relative to species' entire geographic ranges, and individuals encountered in a hybrid zone are probably generations removed from the initial hybridization events. In a 2013 study of a hybrid zone between two different howler monkeys species, Liliana Cortés-Ortiz at the University of Michigan and her then doctoral student Mary Kelaita found that most hybrids in the zone are actually "old" hybrids, meaning that the time when the two species actually interbred was quite a few generations in the past.[50] Since this period, most hybrids had been breeding with one or the other parental species and were found to be very closely similar in appearance to whichever parental species they were breeding with. In a nutshell, the anatomical and outward signs of hybridization seem to be ephemeral. With such insights, the boy from the Old Mill in Portugal and the old woman of the cave in Romania are unlikely to be first-generation hybrids, if they represent hybrids at all.

If they are not hybrids, then how do we explain the mix of Neandertal and modern human features in the boy and in the old woman? One possibility would be that the assumed telltale "hybrid" features of the two different hominin fossils are not definitively telltale. For example, let's look at the occipital bun, the rounded bun-like bump on the back of the skull in the old woman. Recent studies of this feature have shown that when the curvature at the back of the skull in relation to the rest of the cranium is digitally reproduced in 3-D and comparisons are made between modern humans and Neandertals, Neandertals fit within the variation seen in modern humans.[51] Therefore, paleoanthropologists have moved away from pointing to specific bony traits divorced from the rest of the features of the skull and claiming this or that trait is definitely indicative of a Neandertal or a modern human. In other words, the overall anatomical shape of the skull or constellation of features needs to be examined as an integrated structure for these comparisons.

The ear region of the "hybrid" boy's skull is said to have a shape intermediate between modern humans and Neandertals, but Neandertals are known to show significant variation in the shape of this region and it is not possible to get a reliable idea of the adult shape of the ear region from a boy of seven years old. Other claims that the boy had the stocky build of Neandertals suffer from a similar problem: how to judge this reliably on the basis of a young child's skeleton. As has been pointed out, the boy could just as easily have been a well-fed and "chunky" modern human child.[52]

The proof will be in the genomic pudding, so to speak. Our best way of determining whether the boy from the Old Mill in Portugal, the old woman from the cave in Romania, and other fossils that are claimed as hybrids are indeed true hybrids or descendants of hybrids is to extract and study their ancient DNA to see if telltale genomic regions indicate that both Neandertal and modern human DNA are present in the genomes of these individuals. Indeed, extracting their DNA might very well be possible since the ages of the two fossils are very similar to the ages of the fossils from which ancient DNA has already been extracted. If they are indeed actual hybrids, they might be expected to show larger proportions of archaic DNA compared to today's Europeans or Asians, simply because they lived closer to the time of the initial hybrid events. But it should be borne in mind that even if we never find actual hybrids in the fossil record, we now know that archaics and modern humans did interbreed, even if to a minimal degree. Despite the minimal gene exchange between Neandertals and moderns, it should be realized that this gene exchange could conceivably have packed a powerful evolutionary punch. This is especially so if some highly adaptive archaic genes evolved in northern latitudes were quickly spread to colonizing modern humans.

WHAT ELSE CAN ANCIENT DNA TELL US?

For decades, one of the most tantalizing mysteries has been whether the Neandertals spoke, and while fossil bones can tell us nothing on that topic, the genome might indeed have something to say. In chapter 5, we described the *FOXP2* gene that, among other genes, is thought to contribute to human speech and language. This gene, after having been tightly constrained and showing little to no change over many millions of years, suddenly showed two functional DNA changes very recently in the human lineage that all humans have today. Since the question of whether Neandertals could speak has long been hotly debated, we were all holding our breaths with curiosity to see if Neandertals also had these supposedly uniquely human *FOXP2* changes. Thus, in 2007, several years before the full Neandertal genome was determined, Svante Pääbo, along with the Spanish geneticist Carles Lalueza-Fox at the Institute for Evolutionary Biology in Barcelona,[53] made a targeted retrieval of the ancient *FOXP2* gene from the approximately 49,000-year-old bones of two Neandertal individuals from the El Sidrón Cave in northern Spain to check if these Neandertals bore the same two functional changes seen in modern humans. Lo and behold, they did! And, when the Denisovan gene was checked, it also bore the two functional changes in *FOXP2*. This indicated that the two functional changes to *FOXP2* must have occurred at some time previous to the evolutionary separation of humans from the ancestor of Neandertals and Denisovans, perhaps over 400,000 years ago.

The story of *FOXP2* is turning out to be more complex and will require more investigation into how the DNA changes in the gene might have changed its function. While the gene appears to have undergone a beneficial selective sweep in modern humans, indicating there must have been some advantage to the new DNA changes in the gene, the sweep may not be a consequence of the two DNA changes modern humans, Neandertals, and Denisovans all share. Interestingly, in 2013 Pääbo's research group detected a previously unknown DNA change in *FOXP2* that occurred in humans after they had already split from their archaic cousins. They believe this specific DNA change is quite significant, since a change at this precise location in the gene had not occurred in 700 million years of vertebrate evolution.[54] Furthermore, the DNA change was not like the types that had occurred earlier in the gene's evolution, which had changed amino acids in the gene's protein. Instead, Pääbo's laboratory studies on the new DNA change found that it dials down the quantity of protein the gene produces. The emerging story is that the evolution of *FOXP2* was a multiple-step process with important alterations to the protein's structure occurring just before we split from Neandertals and Denisovans (perhaps over 400,000 years ago), and then a subsequent change appearing that down-regulated the gene. (We might call this evolutionary dance through time the *FOXP2*-trot!) Yet, there are many questions still unanswered. For example, while the majority of humans have the DNA change that dials down the gene, about 10% of

Africans do not, yet surely they speak. We need to learn more about these individuals. Meanwhile, though FOXP2 has surely gotten a lot of attention, we must remember that there are certainly multiple other genes that also played important roles in language evolution.

Fossils also do not tell us about their bearer's hair, eye and skin color, or freckles, but the genome might very well tell us. We have learned that many genes play a role in human pigmentation. Recently, researchers have examined more than one hundred DNA variants from different human individuals in over thirty pigmentation genes and arrived at the conclusion that these DNA variants as a group can be used to confidently predict the pigmentation of a person. James Watson, co-discoverer of DNA's structure in 1957, was predicted based on his DNA variants to have blue eyes and brown or blond hair, which indeed he does. Craig Venter, who led the private effort that completed the first full determination of a human genome, was predicted to lack freckles and to have fair skin and blue eyes, which he does. And Henry Louis Gates, an African American professor at Harvard University, was predicted to have no freckles and brown hair, which he does.[55]

Ancient DNA can be like a mirror that reflects what we looked like in the past. When we hold this mirror up to several Neandertals and the single Denisovan, we see varied pigmentation in our archaic cousins. Some individuals show evidence of having light skin; some appear to have had darker skin; some had freckles and others didn't; some are reconstructed to have had blond hair, but others to have red or dark brown hair; and all seem to have had brown eyes. Our ability to predict such traits will become better as we discover new genetic variants and their link to pigmentation, but so far it appears that at least some archaic hominins living in regions outside Africa (where the incidence of ultraviolet light is less intense) were likely to have had fair skin. This would be expected, since these hominins had been living in higher latitudes for up to 400,000 years. They would likely have faced selection pressures to lighten their skin's pigmentation in order to augment vitamin D production, stimulated when sunlight hits the skin, which in turn promotes calcium absorption, helping to keep bones strong.

Interestingly, lightening of the skin in Neandertals and modern humans appears to have evolved convergently in the two, and the same pigmentation gene, MC1R (the melanocortin receptor gene), was involved. Europeans with light skin have DNA alterations in the MC1R gene that reduce the gene's product and lighten the skin and redden the hair. When Lalueza-Fox and Pääbo checked MC1R in 2007 in the two Neandertals from El Sidrón, they found that the gene also contained a DNA alteration that would have reduced the gene's function and which, he believes, indicates that at least some Neandertals had light skin and red hair.[56] But when they screened for the Neandertal DNA alteration in over a thousand modern humans, they could not detect it in any of them, from which they concluded that although the same general mechanism evolved in the two—reducing the function of *MC1R*—the two distant cousins seem to have lightened their skin independent of each other.

We are sure to enjoy many surprises as the field of paleogenetics grows and as genomes from more fossils come to light. Comparisons of the 2014 high-coverage Neandertal genome from southern Siberia with the Denisovan genome have detected very small amounts of Neandertal DNA in the Denisovan genome (on the order of 0.5%), indicating interbreeding occurred between these two extinct hominins and that gene flow was directed from Neandertals into Denisovans. This interbreeding probably

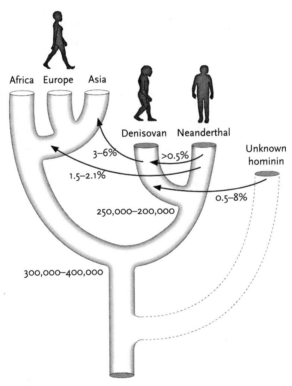

Figure 8.7: An overview of what we have learned about the evolutionary relationships among modern humans and two archaic hominins, Neandertals and Denisovans, by comparing their ancient genomes with those of modern humans. Splitting times, estimated from the genomes, are shown at the places where branches diverge. Thin arrows passing from one hominin lineage to another indicate gene flow due to interbreeding between members of these distinct hominin lineages in the past. Exceedingly ancient DNA in the Denisovan genome raises the strong possibility that it interbred with an ancient hominin lineage that diverged from the human family tree over one million years ago and that presently remains "unknown."

occurred in regions of southern Siberia, since the Denisovan genome is more similar to the south Siberia Neandertal genome compared to Neandertal genomes from Vindija, Croatia, or Mezmaiskaya, Russia. Another very interesting finding is that the fraction of Neandertal DNA in Eurasian people has a higher degree of similarity to the Neandertal genome from Mezmaiskaya (in the Caucasus) than to other Neandertal genomes from southern Siberia or Croatia. This indicates that Neandertals had already established themselves in varied geographic regions in Europe and Asia at the time when interbreeding first took place between modern humans and Neandertal groups.[13]

More surprising is the revelation that the Denisovans likely interbred with a very ancient, and as yet unknown, hominin that diverged from the human family tree sometime over one million years ago, leading to a Denisovan genome riddled with 0.5% to as much as 8% exceptionally ancient DNA (see Figure 8.7). Where this interbreeding took place and who this ancient hominin was remain alluring mysteries. One provocative idea, however, is that this ancient hominin was from a population of *Homo erectus* living in Asia; after all we know *H. erectus* lived in vast regions of Asia over a 1.5 million year period lasting until about 100,000 to 50,000 years ago.[14]

Paleogenomics is unraveling a web of reproductive ties between otherwise distinct ancient hominin populations that inhabited Eurasia. And although the degree of genomic material exchanged appears to have been generally low, it is almost certain these connections will grow ever more complex with new ancient DNA discoveries.

EPILOGUE

Our technology and science have advanced to the stage where we are now firmly in a new scientific age—"the age of genomes." Beyond their undoubted clinical applications, since the first genome was sequenced in 2001, genomes are being used by scientists to examine age-old questions about our evolutionary origins, our unique traits, our evolutionary relationships with other living organisms and—since methods now exist to extract and study ancient DNA and genomes from fossils—our relationships with the extinct cousins with whom we shared this planet for many thousands of years.

Among the significant new discoveries made through analyses of genomic evidence, we have learned without a doubt that chimpanzees (and bonobos) are our closest living relatives among the great apes. We have learned that the diseases of our modern species may be a consequence of the relatively reduced sizes of our ancestors' populations. We have come to realize that some of the high-profile fossils flaunted as our earliest ancestors do not so easily fit within a timeline based on genomes, and might therefore be ancestral doppelgängers—extinct apes near and around the origins of the human lineage who happened to develop similar human features but wound up on evolution's dead-end roads. With respect to our most heralded traits, we have learned that the evolution of our immensely large brains has required the rapid evolution of energy genes that can build and feed such a gluttonous organ. Like voyagers on the sea, we are scanning across genomes and beginning to discover genes that bear signs of natural selection underlying our species' unique features as well as underlying a variety of features in different human populations that allowed them to colonize vastly different environmental regions on Earth.

Inarguably, one of the most significant breakthroughs of the last half dozen years is the recovery of ancient DNA from fossils and its assembly into full genomes of our extinct ancient human cousins, brilliantly lighting up our past. Our analyses of ancient genomes of Neandertals and newly discovered Denisovans has, in one fell swoop, revolutionized human evolutionary studies, banishing any persistent beliefs of our lineage's evolutionary exclusivity. As only an ancient genome can show, we now have direct and indisputable proof that when modern humans dispersed from Africa, we indeed met our long-lost cousins (from over 400,000 years ago) and interbred with them. The almost paradigmatically held view in which modern humans spread out from Africa and replaced our archaic cousins living in Eurasia in one long inexorable march has at long last crumbled, never to stand again.

Such is the power of the new science of human evolution. But beyond a summation of some of the newly unlocked mysteries, the recently discovered answers to lingering questions of anthropologists, I hope to have shown how and why genomes represent such powerful evidence of our past. Their power comes from the very fact that they are composed of many thousands of small independent segments that have been shuffled into various combinations as they were passed down from parents to offspring over the hundreds of thousands of generations of our evolution. If we compare solving human evolution to solving a crime, entire genomes represent the best-preserved crime scenes we could possibly find. Genomes provide literally thousands of pieces of independent evidence that can be gleaned for common patterns pointing to the how, why, and where of our evolutionary past. It's the power of these thousands of little pieces that has finally allowed us to understand conflicting gene trees in the search for our nearest ape cousin; to find true signatures of selection; and to discover that the origin of modern humans was not a temporally and geographically discrete *event* and that, despite the early result that initial mitochondrial genome results suggested, we clearly interbred with our nearest fossil cousins, Neandertals and Denisovans. (Here, it is important to point out that before the determination of the nuclear genome of Neandertals in 2010, mitochondrial DNA sequences extracted from Neandertals during the prior decade, gave us absolutely no hint that we ever interbred with them.)

But the story is far from over. One important aspect of the new science of human evolution is that genomic evidence leads us to ask new questions, some of which we might never have thought to ask. For instance, up until recently we have believed that the adaptations that enabled populations to live in different regions of the world would be marked by a traditional signature known as a selective sweep, a powerful signal confined to single gene regions. Yet, ever since we have started scanning across the genome in search of our adaptations, we have found that by and large these signatures are just not present. This has forced us to rethink how humans have adapted to our environments, which is starting to look like a much more subtle process involving small changes in multiple different regions of our genomes. This realization is influencing the development of new methods designed to detect more subtle signatures of adaptation. Despite these findings, our searches across the genome using traditional methods have actually been successful in turning up lists of genes that possibly underlie important human adaptations. Yet for most of these genes we do not have detailed information about their biological functions. So the new genome evidence is causing us to step back to the laboratory bench in order to discover the detailed functions of genes. This quest will almost certainly lead us to many interesting discoveries about adaptations, but will also help us to better understand the exact functions of genes that were involved in evolving our large brains, our keen cognitive powers, and our complex language abilities.

The revelation that we met and mated with Neandertals and ancient Denisovans raises a myriad of questions. Exactly where did interbreeding take place? Was it restricted only to limited geographic locations, such as the eastern Mediterranean, or did it take place at some level all along the routes that modern humans took as they spread into the territories of our extinct archaic relatives? What can the amounts of archaic genomic bits in different populations tell us about the routes of modern human migrations? And perhaps one of the most important questions of all is whether modern humans imported some beneficial genetic adaptations from our close and recently extinct cousins. Beyond this, the idea that we had close encounters with Neandertals and Denisovans raises renewed questions about how deep and complex our social connections were and how extensive our cultural exchanges were. When we met, to what extent did we share our different knowledge bases—our different technologies and belief systems? And, what ultimately led to their demise and disappearance?

And, since we now have proof of our interbreeding with Neandertals and Denisovans, it seems to make sense that we begin to ask the question: which other extinct members of the human family tree did we possibly interbreed with during our evolutionary journey to becoming modern humans and how might these extinct relatives have contributed to our genome today? For example, there is evidence that *Homo erectus* populations in regions of East Asia (China) and Southeast Asia (Java) may have existed until very recently,[1] making it a distinct possibility that modern humans met and mated with them when we first colonized these regions. There is also the possibility that other as yet undiscovered archaic cousins will come to light (as was the case with Denisovans) and that we will find evidence that we also interbred with them. The discovery in 2014 of exceedingly ancient genomic material in the Denisovan genome, pointing to its liason with a million-year-old hominin, already indicates to us that the reproductive ties among members of the different branches of the human evolutionary tree were far more complex than we had ever thought.

Today we have developed a detailed appreciation of our place in the primate evolutionary tree. This knowledge provides the essential evolutionary framework by which we can trace our adaptations back in time to learn when in primate history these adaptations first started to take shape. While certain human features will likely prove to have uniquely evolved along the human lineage, many others presumed to be unique will be found to have deeper origins within our shared past with other primates. We already know that several genes important in the evolution of our increased brain size also show signatures of adaptation on the common ancestral stem we share with apes. Furthermore, we are learning that even for features in which we thought we were unique, such as our large brains and long lifespans, it appears we have evolved in remarkably similar genetic ways to even distantly related animals like elephants and bottlenose dolphins. Even in our complex language abilities, we share some of the same genetic substrates as other mammals and even songbirds. Our continued quest to better understand our genome and evolutionary past will almost certainly reveal the many ways we are deeply connected to the rest of the animal world. As Charles Darwin himself suggested, "the difference in mind between man and the higher animals, great as it is, certainly is one of degree and not of kind."

Indeed, our quest began over 150 years ago when Darwin wrote in *The Origin of Species* the modest statement that "light will be thrown on the origin of man and his history." If Darwin were alive today, he would bear witness to important clues that researchers almost weekly discover in our genomes that illuminate ever more brightly our evolutionary past.

GLOSSARY

1000 Genomes Project—started in 2008, aims to determine the complete DNA sequence of entire genomes from well over a thousand people from different world populations.

Allele—is one out of the two or more versions of a particular gene, usually differing by one or more DNA mutations.

Allopatric speciation—the divergence of two populations due to reproductive isolation following separation of the two by a geographic barrier.

Autosomal DNA—all DNA from chromosomes in the nucleus except from the sex chromosomes.

Chromosome—a long strand of coiled DNA found inside the nucleus.

Coalescence—proceeding back into evolutionary time, the confluence of two gene lineages into a single common ancestral gene.

Divergence—the gradual accumulation of DNA differences between two different populations or species since they split from their most recent common ancestor.

DNA variant—is any difference in the DNA between two people. It can be at a single DNA site, or it can involve more than one DNA site. SNPs (or differences at single DNA sites) are examples of DNA variants.

Effective population size—is an idealized measure of the average size of a population based on the number of individuals that contribute genes to the next generation. Because of variance in the reproduction of different individuals it will almost always be less than the census size, often considerably so.

ENCODE project—(The Encyclopedia of DNA Elements) is an international project initiated in 2003 to determine all functional elements in the human genome.

Epistasis—an interaction in which a DNA change at one site in the genome affects the phenotypic effect of a DNA change at another site in the genome.

Gene—a segment of heritable DNA that codes for a protein.

Gene flow—the passage of genetic information from one population to another through interbreeding.

Gene surfing—when gene versions that happen to be present at the frontier of a spatially expanding population, even if rare, ride the wave of advance and become quite common in the population after the expansion.

Gene tree—a tree diagram that shows the evolutionary relationships among gene copies or segments of the genome (often these copies are from different species).

Genetic drift (Drift)—random changes in the frequency of a DNA variant or allele in a population (i.e., how common it is in the population or species) as a result of random variation in individual reproduction.

HAPMAP Project (Haplotype Map of genetic variations)—started in 2002, is an international effort aimed to identify and catalog genetic similarities and differences in diverse humans. The project screens for already determined SNPs and for their presence in other populations.

Homology—features in organisms that were inherited from their shared common ancestor. Also, genes in different species, present at the same locus or position on a chromosome, which were inherited from a common ancestor.

Homoplasy—similar features evolving in unrelated organisms; non-homologous features.

Linkage group—DNA bases found close to each other on the same chromosome are often inherited as a group, and are linked together in inheritance.

Mitochondrial DNA (mtDNA)—is the DNA contained in organelles called mitochondria inside animal cells. In humans, it is a circular genome of 16,500 nucleotide bases organized into thirty-seven genes. It is inherited through the maternal lineage only.

Modern human—is a term often equated with a member of the species *Homo sapiens*, but can be used to also designate an ancient human with some or mostly modern human anatomical features.

Natural selection—the differential survival or reproduction of individuals based on their fitness within the specific environment in which they live.

Negative selection—a form of natural selection that removes deleterious DNA variants or alleles leading these variants to become less common or entirely absent in the population. Also called purifying selection.

Nuclear DNA—is DNA contained in the nucleus of animal cells. It is organized into twenty-three chromosomes, including one pair of sex chromosomes. Its inheritance is through both parents.

Parapatric speciation—Geographically contiguous populations of a widespread species will meet at zones where they show reduced or limited interbreeding. Over time the two populations can become reproductively isolated even in the midst of continued but slowing gene flow and become different species.

Perlegen project—was a private initiative by Perlegen Sciences, Inc. to catalog genetic similarities and differences in people from different global populations. The project screened for already ascertained SNPs to check for their presence or absence in other populations.

Phenotype—the observable characteristics of an organism.

Positive selection—a form of natural selection that favors one DNA variant or allele over another and thereby leads them to become more common in the population.

Random lineage sorting—When different versions of a gene exist in the common ancestor of three species, it is possible that each of the three copies sorts randomly through evolutionary time into the lineages of the three species. This results in a fraction of gene trees that will not match the species tree.

Selective sweep—As a DNA copy bearing a beneficial DNA variant spreads through a population and replaces all other DNA copies, the DNA variants found on those other copies are "swept" away.

SNP—stands for Single Nucleotide Polymorphism and is pronounced as "Snips." It refers to sites along a DNA sequence where the nucleotide bases (A, C, G, or T) between two different people or between paired chromosomes are different (i.e., are DNA variants).

Speciation—the process whereby one species splits to form two or more reproductively isolated species.

Species—a population whose members interbreed under natural conditions, produce fertile offspring, and are reproductively isolated from other such groups.

Species tree—In contrast to a gene tree, it is a tree diagram representing the evolutionary relationships among species.

Substitution—a DNA difference between two different species.

NOTES

PROLOGUE

1. Lindblad-Toh, K., et al. A high-resolution map of human evolutionary constraint using 29 mammals. *Nature* **478**(7370):476–482 (2011).
2. Ponting, C. P., and Hardison, R.C. What fraction of the human genome is functional? *Genome Res.* **21**(11):1769–1776 (2011).
3. Darwin, C. *The Descent of man and selection in relation to sex.* (London: John Murray; 1871).

CHAPTER 1

1. Patterson, C. *Molecules and morphology in evolution: Conflict or compromise?* (New York: Cambridge University Press; 1987).
2. Shipman, P., et al. Butchering of giant geladas at an Acheulian site. *Curr. Anthropol.* **22**(3):257–268 (1981).
3. Disotell, T. R., Honeycutt, R. L., and Ruvolo, M. Mitochondrial DNA phylogeny of the old-world monkey tribe *Papionini*. *Mol. Biol. Evol.* **9**(1):1–13 (1992).
4. Young, N. M. A reassessment of living hominoid postcranial variability: implications for ape evolution. *J. Hum. Evol.* **45**(6):441–464 (2003).
5. Manceau, M., et al. Convergence in pigmentation at multiple levels: mutations, genes and function. *Philos. Trans. R. Soc. Lond. Series B. Biol. Sci.* **365**(1552):2439–2450 (2010).
6. Shapiro, M. D., Bell, M., and Kingsley, D. M. Parallel genetic origins of pelvic reduction in vertebrates. *Proc. Natl. Acad. Sci. U. S. A.* **103**(137):13753–13758 (2006).
7. Protas, M. E., et al. Genetic analysis of cavefish reveals molecular convergence in the evolution of albinism. *Nat. Genet.* **38**(1):107–111 (2005).
8. Arendt, J., and Reznick, D. Convergence and parallelism reconsidered: What have we learned about the genetics of adaptation? *Trends Ecol. Evol.* **23**(1):26–32 (2008).
9. Collard, M., and Wood, B. How reliable are human phylogenetic hypotheses? *Proc. Natl. Acad. Sci. U.S.A.* **97**(9):5003–5006 (2000).
10. Collard, M., and Wood, B. Homoplasy and the early hominid masticatory system: inferences from analyses of extant hominoids and papionins. *J. Hum. Evol.* **41**(3):167–194 (2001).
11. Fleagle, J. G., and McGraw, W. S. Skeletal and dental morphology of African papionins: unmasking a cryptic clade. *J. Hum. Evol.* **42**(3):267–292 (2002).
12. Gilbert, C. C., Frost, S. R., and Strait, D. S. Allometry, sexual dimorphism, and phylogeny: a cladistic analysis of extant African papionins using craniodental data. *J. Hum. Evol.* **57**(3):298–320 (2009).

13. Wood, B. Colloquium paper: reconstructing human evolution: achievements, challenges, and opportunities. *Proc. Natl. Acad. Sci. U.S.A.* **107**(Suppl 2):8902–8909 (2010).

14. Wood, B. Paleoanthropology: Facing up to complexity. *Nature* **488**(7410):162–163 (2012).

CHAPTER 2

1. Goodman, M. Serological analysis of the systematics of recent hominoids. *Hum. Biol.* **35**(3):377–436 (1963).

2. Sarich, V., and Wilson, A. Quantitative immunochemistry and the evolution of primate albumins: micro-complement fixation. *Science* **154**(3756):1563–1566 (1966).

3. Darwin, C. *The Descent of man and selection in relation to sex.* (London: John Murray; 1871).

4. Sibley, C., and Ahlquist, J. The phylogeny of the hominoid primates, as indicated by DNA-DNA hybridization. *J. Mol. Evol.* **20**(1):2–15 (1984).

5. Galili, U., and Swanson, K. Gene sequences suggest inactivation of alpha-1,3-galactosyltransferase in catarrhines after the divergence of apes from monkeys. *Proc. Natl. Acad. Sci. U. S. A.* **88**(16):7401–7404 (1991).

6. Djian, P., and Green, H. Vectorial expansion of the involucrin gene and the relatedness of the hominoids. *Proc. Natl. Acad. Sci. U. S. A.* **86**(21):8447–8451 (1989).

7. Kawamura, S., et al. Evolutionary rate of immunoglobulin alpha noncoding region is greater in hominoids than in Old World monkeys. *Mol. Biol. Evol.* **8**(6):743–752 (1991).

8. Ruvolo, M., et al. Resolution of the African hominoid trichotomy by use of a mitochondrial gene sequence. *Proc. Natl. Acad. Sci. U. S. A.* **88**(4):1570–1574 (1991).

9. Bailey, W. J., et al. Reexamination of the African hominoid trichotomy with additional sequences from the primate beta-globin gene cluster. *Mol. Phylogenet. Evol.* **1**(2):97–135 (1992).

10. Retief, J., et al. Evolution of protamine P1 genes in primates. *J. Mol. Evol.* **37**(4):426–434 (1993).

11. Retief, J., and Dixon, G. Evolution of pro-protamine P2 genes in primates. *Eur. J. Biochem.* **214**(2):609–615 (1993).

12. Pamilo, P., and Nei, M. Relationships between gene trees and species trees. *Mol. Biol. Evol.* **5**(5):568–583 (1988).

13. Degnan, J. H., and Rosenberg, N. A. Gene tree discordance, phylogenetic inference and the multispecies coalescent. *Cell* **24**(6):332–340 (2009).

14. Stewart, C.-B., Schilling, J. and Wilson, A. Adaptive evolution in the stomach lysozymes of foregut fermenters. *Nature* **330**,(6146):401–404 (1987).

15. Ruvolo, M. Molecular phylogeny of the hominoids: Inferences from multiple independent DNA sequence data sets. **14**(3):248–265 (1997).

16. Wu, Chung-I. Inferences of species phylogeny in relation to segregation of ancestral polymorphisms. *Genetics* **127**(2):429–435 (1991).

17. Chimpanzee Sequencing and Analysis Consortium. Initial sequence of the chimpanzee genome and comparison with the human genome. *Nature* **437**(7055):69–87 (2005).

18. Prüfer, K., et al. The bonobo genome compared with the chimpanzee and human genomes. *Nature* **486**(7404):527–531 (2012).

19. Prado-Martinez, J., et al. Great ape genetic diversity and population history. *Nature* **499**(7459):471–475 (2013).

20. Maddison, W. P. Gene trees in species trees. *Syst. Biol.* **46**(3):523–536 (1997).

21. Hey, J. *Genes, categories, and species.* (Oxford: Oxford University Press; 2001).

22. Patterson, N., et al. Genetic evidence for complex speciation of humans and chimpanzees. *Nature* **441**(7097):1103–1108 (2006).

23. Ebersberger, I., et al. Mapping human genetic ancestry. *Mol. Biol. Evol.* **24**(10):2266–2276 (2007).

24. Kappelman, J., et al. The earliest occurrence of *Sivapithecus* from the middle Miocene Chinji Formation of Pakistan. *J. Hum. Evol.* **21**(1):61–73 (1991).

25. Hobolth, A., et al. Incomplete lineage sorting patterns among human, chimpanzee, and orangutan suggest recent orangutan speciation and widespread selection. *Genome Res.* **21**(3):349–356 (2011).

CHAPTER 3

1. Relethford, J. *Genetics and the search for modern human origins.* (New York: Wiley-Liss; 2001).

2. Gronau, I., et al. Bayesian inference of ancient human demography from individual genome sequences. *Nat. Genet.* **43**(10):1031–1034 (2012).

3. Marth, G., et al. Sequence variations in the public human genome data reflect a bottlenecked population history. *Proc. Natl. Acad. Sci. U. S. A.* **100**(1):376–381 (2003).

4. Bocquet-Appel, J.-P. When the world's population took off: the springboard of the Neolithic Demographic Transition. *Science* **333**(6042):560–561 (2011).

5. Gignoux, C. R., Henn, B. M., and Mountain, J. L. Rapid, global demographic expansions after the origins of agriculture. *Proc. Natl. Acad. Sci. U. S. A.* **108**(15):6044–6049 (2011).

6. Burgess, R., and Yang, Z. Estimation of hominoid ancestral population sizes under Bayesian coalescent models incorporating mutation rate variation and sequencing errors. *Mol. Biol. Evol.* **25**(9):1979–1994 (2008).

7. Pamilo, P., and Nei, M. Relationships between gene trees and species trees. *Mol. Biol. Evol.* **5**(5):568–583 (1988).

8. Rogers, J. The phylogenetic relationships among Homo, Pan, and Gorilla: a population genetics perspective. *J. Hum. Evol.* **25**(3):201–215 (1993).

9. Ruvolo, M. Molecular phylogeny of the hominoids: Inferences from multiple independent DNA sequence data sets. **14**(3):248–265 (1997).

10. Chen, F. C., and Li, W. H. Genomic divergences between humans and other hominoids and the effective population size of the common ancestor of humans and chimpanzees. *Am. J. Hum. Genet.* **68**(2):444–456 (2001).

11. Kappelman, J., et al. The earliest occurrence of *Sivapithecus* from the middle Miocene Chinji Formation of Pakistan. *J. Hum. Evol.* **21**(1):61–73 (1991).

12. Chimpanzee Sequencing and Analysis Consortium. Initial sequence of the chimpanzee genome and comparison with the human genome. *Nature* **437**(7055):69–87 (2005).

13. McVicker, G., et al. Widespread genomic signatures of natural selection in hominid evolution. *PLoS Genet.* **5**(5):e1000471 (2009).

14. Hobolth, A., et al. Incomplete lineage sorting patterns among human, chimpanzee, and orangutan suggest recent orangutan speciation and widespread selection. *Genome Res.* **21**(3):349–356 (2011).

15. Scally, A., and Durbin, R. Revising the human mutation rate: implications for understanding human evolution. *Nat. Rev. Genet.* **13**(10):745–753 (2012).

16. Hobolth, A., et al. Genomic relationships and speciation times of human, chimpanzee, and gorilla inferred from a coalescent hidden Markov model. *PLoS Genet.* **3**(2):e7 (2007).

17. Patterson, N., et al. Genetic evidence for complex speciation of humans and chimpanzees. *Nature* **441**(7097):1103–1108 (2006).

18. Scally, A., et al. Insights into hominid evolution from the gorilla genome sequence. *Nature* **483**(7388):169–175 (2012).

19. Mayr, E. *Animal species and evolution.* (Cambridge: Harvard Univ. Press; 1963).

20. Mayr, E. *Systematics and the origin of species.* (New York: Colombia Univ. Press; 1942).

21. Jolly, C.J. A proper study for mankind: analogies from the papionin monkeys and their implications for human evolution. *Am. J. Phys. Anthropol.* **116**(Suppl 33):177–204(2001).

22. Tosi, A.J., et al. Cercopithecine Y-chromosome data provide a test of competing morphological evolutionary hypotheses. *Mol. Phylogenet. Evol.* **27**(3):510–521 (2003).

23. Hey, J. *Genes, categories, and species.* (New York: Oxford University Press; 2001).

24. Barton, N. Evolutionary biology: How did the human species form? *Curr. Biol.* **16**(16):647–650 (2006).

25. Disotell, T. R. "Chumanzee" evolution: the urge to diverge and merge. *Genome Biol.* **7**(11):240 (2006).

26. Wakeley, J. Complex speciation of humans and chimpanzees. *Nature* **452**(7184):E3–4, discussion E4 (2008).

27. Siepel, A. Phylogenomics of primates and their ancestral populations. *Genome Res.* **19**(11):1929–1941 (2009).

28. Yamamichi, M., Gojobori, J., and Innan, H. An autosomal analysis gives no genetic evidence for complex speciation of humans and chimpanzees. *Mol. Biol. Evol.* **29**(1):145–156 (2012).

29. Zhu, T., and Yang, Z. Maximum likelihood implementation of an isolation-with-migration model with three species for testing speciation with gene flow. *Mol. Biol. Evol.* **29**(10):3131–3142 (2012).

30. Mailund, T., et al. A new isolation with migration model along complete genomes infers very different divergence processes among closely related great ape species. *PLoS Genet.* **8**(12):e1003125 (2012).

31. Prüfer, K., et al. The bonobo genome compared with the chimpanzee and human genomes. *Nature* **486**(7404):527–531 (2012).

32. Harrison, T. Apes among the tangled branches of human origins. *Science* **327**(5965):532–534 (2010).

33. Wood, B., and Harrison, T. The evolutionary context of the first hominins. *Nature* **470**(7334):347–352 (2011).

34. Yang, Z. A likelihood ratio test of speciation with gene flow using genomic sequence data. *Genome Biol. Evol.* **2**(1):200–211 (2010).

35. Prado-Martinez, J., et al. Great ape genetic diversity and population history. *Nature* **499**(7459):471–475 (2013).

36. Ebersberger, I., et al. Mapping human genetic ancestry. *Mol. Biol. Evol.* **24**(10):2266–2276 (2007).

CHAPTER 4

1. Lieberman, D. *The Story of the Human Body: Evolution, Health, and Disease* (New York: Pantheon Books; 2013).

2. Cann, R., Stoneking, M., and Wilson, A. Mitochondrial DNA and human evolution. *Nature* **325**(6099):31–36 (1987).

3. Vigilant, L., et al. African populations and the evolution of human mitochondrial DNA. *Science* **253**(5027):1503–1507 (1991).

4. Harris, E. E., and Hey, J. X chromosome evidence for ancient human histories. *Proc. Natl. Acad. Sci. U. S. A.* **96**(6):3320–3324 (1999).

5. Jaruzelska, J., et al. Spatial and temporal distribution of the neutral polymorphisms in the last ZFX intron: analysis of the haplotype structure and genealogy. *Genetics* **152**(3):1091–1101 (1999).

6. Garrigan, D., et al. Deep haplotype divergence and long-range linkage disequilibrium at xp21.1 provide evidence that humans descend from a structured ancestral population. *Genetics* **170**(4):1849–1856 (2005).

7. Harris, E. E., and Hey, J. Human populations show reduced DNA sequence variation at the Factor IX locus. *Curr. Biol.* **11**(10):774–778 (2001).

8. Blum, M. G. B., and Jakobsson, M. Deep divergences of human gene trees and models of human origins. *Mol. Bio. Evol.* **28**(2)889–898 (2011).

9. Gronau, I., et al. Bayesian inference of ancient human demography from individual genome sequences. *Nat. Genet.* **43**(10):1031–1034 (2011).

10. Arcos-Burgos, M., and Muenke, M. Genetics of population isolates. *Clin. Genet.* **61**(4):233–247 (2002).

11. Sacks, O. *The island of the colorblind and Cycad Island.* (New York: Alfred A. Knopf; 1997).

12. Ben Simon, G. J. Pingelapese achromatopsia: correlation between paradoxical pupillary response and clinical features. *Br. J. Ophthalmol.* **88**(2):223–225 (2004).

13. Mantle, J., and Pepys, J. Asthma amongst Tristan da Cunha islanders. *Clin. Allergy* **4**(2):161–170 (1974).

14. Thompson, M., et al. A functional G300S variant of the cysteinyl leukotriene 1 receptor is associated with atopy in a Tristan da Cunha isolate. *Pharmacogenet. Genomics* **17**(7):539–549 (2007).

15. Getahun, D., Demissie, K., and Rhoads, G. Recent trends in asthma hospitalization and mortality in the United States. *J. Asthma* **42**(5):373–378 (2005).

16. Glass, B., et al. Genetic drift in a religious isolate, an analysis of the causes of variation in blood group and other gene frequencies in a small population. *Am. Nat.* **86**(828):145–159 (1952).

17. Lamont, R., et al. A locus for Bowen-Conradi syndrome maps to chromosome region 12p13.3. *Am. J. Med. Genet.* **132A**(2):136–143 (2005).

18. Dugoff, L., Thieme, G., and Hobbins, J. First trimester prenatal diagnosis of chondroectodermal dysplasia (Ellis-van Creveld syndrome) with ultrasound. *Ultrasound Obstet. Gynecol.* **17**(1):86–88 (2001).

19. Rozenberg, R., and Pereira, L. da Vega. The frequency of Tay-Sachs disease causing mutations in the Brazilian Jewish population justifies a carrier screening program. *São Paulo Med. J.* **119**(4):146–149 (2001).

20. Woolf, C. Albinism (OCA2) in Amerindians. *Am. J. Phys. Anthropol.* Suppl 41:118–140 (2005).

21. Friedlaender, J. S., et al. The genetic structure of Pacific Islanders. *PLoS Genet.* **4**(1):e19 (2008).

22. Norton, H. L., et al. Skin and hair pigmentation variation in island Melanesia. *Am. J. Phys. Anthropol.* **130**(2):254–268 (2006).

23. Kosiol, C., et al. Patterns of positive selection in six mammalian genomes. *PLoS Genet.* **4**(8):e1000144 (2008).

24. Keightley, P. D., Lercher, M. J., and Eyre-Walker, A. Evidence for widespread degradation of gene control regions in hominid genomes. *PLoS Biol.* **3**(2):e42 (2005).

25. Gibbs, R. A., et al. Evolutionary and biomedical insights from the rhesus macaque genome. *Science* **316**(5822):222–234 (2007).

26. Hughes, A. L., and Friedman, R. More radical amino acid replacements in primates than in rodents: support for the evolutionary role of effective population size. *Gene* **440**(1–2):50–6 (2009).

27. Abecasis, G. R., et al. An integrated map of genetic variation from 1,092 human genomes. *Nature* **491**(7422):56–65 (2012).

28. Tennessen, J. A., et al. Evolution and functional impact of rare coding variation from deep sequencing of human exomes. *Science* **337**(6090):64–69 (2012).

29. Manolio, T. A., et al. Finding the missing heritability of complex diseases. *Nature* **461**(7265):747–753 (2009).

30. Wang, X., Grus, W. E., and Zhang, J. Gene losses during human origins. *PLoS Biol.* **4**(3):e52 (2006).

31. Varki, A., and Altheide, T. K. Comparing the human and chimpanzee genomes: searching for needles in a haystack. *Genome Res.* **15**(12):1746–1758 (2005).

32. Olson, M. V. Molecular evolution '99—when less is more: Gene loss as an engine of evolutionary change. *Am. J. Hum. Genet.* 64(1):18–23 (1999).

33. Stedman, H. H., et al. Myosin gene mutation correlates with anatomical changes in the human lineage. *Nature* **428**(6981):415–418 (2004).

34. Kondrashov, A. S., Sunyaev, S., and Kondrashov, F. A. Dobzhansky–Muller incompatibilities in protein evolution. *Proc. Natl. Acad. Sci. U. S. A.* 99(23):14878–14883 (2002).

35. Plotnikova, O. V., et al. Conversion and compensatory evolution of the gamma-crystallin genes and identification of a cataractogenic mutation that reverses the sequence of the human CRYGD gene to an ancestral state. *Am. J. Hum. Genet.* **81**(1):32–43 (2007).

36. Scally, A., et al. Insights into hominid evolution from the gorilla genome sequence. *Nature* **483**(7388)169–175 (2012).

37. Ye, B., et al. A common human SCN5A polymorphism modifies expression of an arrhythmia causing mutation. *Physiol. Genomics* **12**(3):187–193 (2003).

38. Warnecke, T., and Rocha, E. P. C. Function-specific accelerations in rates of sequence evolution suggest predictable epistatic responses to reduced effective population size. *Mol. Biol. Evol.* **28**(8):2339–2349 (2011).

39. Pritchard, J. K., Pickrell, J. K., and Coop, G. The genetics of human adaptation: hard sweeps, soft sweeps, and polygenic adaptation. *Curr. Biol.* **20**(4):R208–215 (2010).

40. Harris, E. E. Nonadaptive processes in primate and human evolution. *Am. J. Phys. Anthropol.* **143**(Suppl 51):13–45 (2010).

41. Eyre-Walker, A. The genomic rate of adaptive evolution. *Trends Ecol. Evol.* **21**(10):569–575 (2006).

42. Hawks, J., et al. Recent acceleration of human adaptive evolution. *Proc. Natl. Acad. Sci. U.S.A.* **104**(2):20753–20758 (2007).

CHAPTER 5

1. Carroll, L. *Through the looking glass: And what Alice found there* (Philadelphia, PA: Henry Altemus Company; 1897).

2. Wong, A. Testing the effects of mating system variation on rates of molecular evolution in primates. *Evolution* **64**(9):2779–2785 (2010).

3. Dixson, A. F. *Primate sexuality: Comparative studies of the prosimians, monkeys, apes and human beings* (New York: Oxford University Press; 1998).

4. Rightmire, G. P. *The evolution of Homo erectus: Comparative anatomical studies of an extinct human species* (New York: Cambridge University Press; 1990).

5. Hughes, A. L. Looking for Darwin in all the wrong places: The misguided quest for positive selection at the nucleotide sequence level. *Heredity (Edinb)*. **99**(4):364–73 (2007).

6. Chimpanzee Sequencing and Analysis Consortium. Initial sequence of the chimpanzee genome and comparison with the human genome. *Nature* **437**(7055):69–87 (2005).

7. Roth, G., and Dicke, U. Evolution of the brain and intelligence. *Trends Cogn. Sci.* **9**(5):250–257 (2005).

8. Wang, H.-Y., et al. Rate of evolution in brain-expressed genes in humans and other primates. *PLoS Biol.* **5**(2):e13 (2007).

9. Wang, J., Li, Y., and Su, B. A common SNP of MCPH1 is associated with cranial volume variation in Chinese population. *Hum. Mol. Genet.* **17**(9):1329–1335 (2008).

10. Rimol, L. M., et al. Sex-dependent association of common variants of microcephaly genes with brain structure. *Proc. Natl. Acad. Sci. U.S.A.* **17**(1):384–388 (2010).

11. Montgomery, S. H., et al. Adaptive evolution of four microcephaly genes and the evolution of brain size in anthropoid primates. *Mol. Biol. Evol.* **28**(1):625–638 (2011).

12. Leonard, W., Snodgrass, J., and Robertson, M. Effects of brain evolution on human nutrition and metabolism. *Annu. Rev. Nutr.* **27**(1):311–327 (2007).

13. Grossman, L. I., et al. Accelerated evolution of the electron transport chain in anthropoid primates. *Trends Genet.* **20**(11):578–585 (2004).

14. Aiello, L. C., Wheeler, P., and Chivers, D. The expensive-tissue hypothesis: The brain and the digestive system in human and primate evolution. *Curr. Anthropol.* **36**(2):199–221 (2012).

15. Fedrigo, O., et al. A potential role for glucose transporters in the evolution of human brain size. *Brain. Behav. Evol.* **78**(4):315–26 (2011).

16. Uddin, M., et al. Distinct genomic signatures of adaptation in pre- and postnatal environments during human evolution. *Proc. Natl. Acad. Sci. U.S.A.* **105**(9):3215–3220 (2008).

17. Goodman, M., et al. Phylogenomic analyses reveal convergent patterns of adaptive evolution in elephant and human ancestries. *Proc. Natl. Acad. Sci. U.S.A.* **106**(49):20824–20829 (2009).

18. McGowen, M. R., Grossman, L. I., and Wildman, D. E. Dolphin genome provides evidence for adaptive evolution of nervous system genes and a molecular rate slowdown. *Proc. Biol. Sci.* **279**(1743):3643–3651 (2012).

19. Wrangham, R. W. *Catching fire: How cooking made us human.* (New York: Basic Books; 2009).

20. Stedman, H. H., et al. Myosin gene mutation correlates with anatomical changes in the human lineage. *Nature* **428**(6981):415–418 (2004).

21. Pfefferle, A. D., et al. Comparative expression analysis of the phosphocreatine circuit in extant primates: Implications for human brain evolution. *J. Hum. Evol.* **60**(2):205–212 (2011).

22. Finch, C. E., and Stanford, C. Meat-adaptive genes and the evolution of slower aging in humans. *Q. Rev. Biol.* **79**(1):3–50 (1984).

23. Sapolsky, R., and Finch, C. Alzheimer's disease and some speculations about the evolution of its modifiers. *Ann. N. Y. Acad. Sci.* **924**:99–103 (2000).

24. King, M. and Wilson, A. C. Evolution at two levels in humans and chimpanzees, *Science*. **188**(4184):107–116 (2007).

25. Ohno, S. So much "junk" DNA in our genome. *Brookhaven Symp. Biol.* **23**:366–370 (1972).

26. Lindblad-Toh K., et al. A high-resolution map of human evolutionary constraint using 29 mammals. *Nature* **478**(7370):476–482 (2011).

27. Enattah, N. S., et al. Identification of a variant associated with adult-type hypolactasia. *Nat. Genet.* **30**(2):233–237 (2002).

28. Cáceres, M., et al. Elevated gene expression levels distinguish human from non-human primate brains. *Proc. Natl. Acad. Sci. U.S.A.* **100**(22):13030–13035 (2003).

29. Konopka, G., et al. Human-specific transcriptional networks in the brain. *Neuron* **75**(4):601–617 (2012).

30. Somel, M., et al. Transcriptional neoteny in the human brain. *Proc. Natl. Acad. Sci. U.S.A.* **106**(14):5743–5748 (2009).

31. Gould, S. J. *Ontogeny and phylogeny.* (Cambridge, MA: Harvard University Press; 1977).

32. Miller, D. J., et al. Prolonged myelination in human neocortical evolution. *Proc. Natl. Acad. Sci. U.S.A.* **109**(41):16480–16485 (2012).

33. Morris, D. *The naked ape: A zoologist's study of the human animal.* (London: Cape; 1967).

34. Rees, J. L. and Harding, R. M. Understanding the evolution of human pigmentation: recent contributions from population genetics. *J. Invest. Dermatol.* **132**(3):846–853 (2012).

35. Harding, R. M., et al. Evidence for variable selective pressures at MC1R. *Am. J. Hum. Genet.* **66**(4):1351–1361 (2000).

36. Rogers, A. R., Iltis, D., and Wooding, S. Genetic variation at the MC1R locus and the time since loss of human body hair. *Curr. Anthropol.* **45**(1):105–108 (2004).

37. Reed, D. L., et al. Pair of lice lost or parasites regained: the evolutionary history of anthropoid primate lice. *BMC Biol.* **5**(1):7 (2007).

38. Jablonski, N. G., and Chaplin, G. The evolution of human skin coloration. *J. Hum. Evol.* **39**(1):57–106 (2000).

39. Falk, D. Brain evolution in *Homo*: The "radiator" theory. *Behav. Brain Sci.* **13**(2):333–381 (1990).

40. Bramble, D.M., and Lieberman, D.E. Endurance running and the evolution of *Homo. Nature* **432**(7015):345–52 (2004).

41. Klein, R. G. Archeology and the evolution of human behavior. *Evol. Anthropol.* **9**(1):17–36 (2000).

42. McBrearty, S., and Brooks, A.S. The revolution that wasn't: a new interpretation of the origin of modern human behavior. *J. Hum. Evol.* **39**(5):453–563 (2000).

43. Hickok, G., and Poeppel, D. The cortical organization of speech processing. *Nat. Rev. Neurosci.* **8(5)**:393–402 (2007).

44. Enard, W., et al. A humanized version of Foxp2 affects cortico-basal ganglia circuits in mice. *Cell* **137**(5):961–971 (2009).

45. Bolhuis, J. J., Okanoya, K. and Scharff, C. Twitter evolution: converging mechanisms in birdsong and human speech. *Nat. Rev. Neurosci.* **11**(11):747–759 (2010).

46. Preuss, T. M. Human brain evolution: From gene discovery to phenotype discovery. *Proc. Natl. Acad. Sci. U.S.A.* **109**(Suppl 1):10709–10716 (2012).

47. Clark, A. G., et al. Inferring nonneutral evolution from human-chimp-mouse orthologous gene trios. *Science* **302**(5652):1960–1963 (2003).

48. Wong, P. C. M., Chandrasekaran, B., and Zheng, J. The derived allele of ASPM is associated with lexical tone perception. *PLoS One* **7**(4):e34243 (2012).

49. Higham, T., et al. Testing models for the beginnings of the Aurignacian and the advent of figurative art and music: the radiocarbon chronology of Geißenklösterle. *J. Hum. Evol.* **62**(6):664–676 (2012).

50. Chan, Y. F., et al. Adaptive evolution of pelvic reduction in sticklebacks by recurrent deletion of a Pitx1 enhancer. *Science* **327**(5963):302–305 (2010).

51. Abzhanov, A., et al. The calmodulin pathway and evolution of elongated beak morphology in Darwin's finches. *Nature* **442**(7102):563–567 (2006).

52. Khan, Z., et al. Primate transcript and protein expression levels evolve under compensatory selection pressures. *Science* **342**(6162):1100–1104 (2013).

CHAPTER 6

1. Vigilant, L., et al. African populations and the evolution of human mitochondrial DNA. *Science* **253**(5027):1503–1507 (1991).

2. Ingman, M., Kaessmann, H., Pääbo, S., and Gyllensten U. Mitochondrial genome variation and the origin of modern humans. **408**(6813):708–713 (2000).

3. Behar, D. M., et al. A "Copernican" reassessment of the human mitochondrial DNA tree from its root. *Am. J. Hum. Genet.* **90**(4):675–684 (2012).

4. Thomson, R., et al. A recent common ancestry of human Y chromosomes: Evidence from DNA sequence data. *Nature.* **378**(6555):376–378 (1995).

5. Ramachandran, S., et al. Support from the relationship of genetic and geographic distance in human populations for a serial founder effect originating in Africa. *Proc. Natl. Acad. Sci. U. S. A.* **102**(44):15942–15947 (2005).

6. Betti, L., et al. The relative role of drift and selection in shaping the human skull. *Am. J. Phys. Anthropol.* **141**(1):76–82 (2010).

7. Betti, L., von Cramon-Taubadel, N., and Lycett, S. J. Human pelvis and long bones reveal differential preservation of ancient population history and migration out of Africa. *Hum. Biol.* **84**(2):139–152 (2012).

8. Moodley, Y., et al. Age of the association between *Helicobacter pylori* and man. *PLoS Pathog.* **8**(5):e1002693 (2012).

9. Tanabe, K., et al. *Plasmodium falciparum* accompanied the human expansion out of Africa. *Curr. Biol.* **20**(14):1283–1289 (2010).

10. Henn, B. M., et al. Hunter-gatherer genomic diversity suggests a southern African origin for modern humans. **108**(13):5154–5162 (2011).

11. Lachance, J., et al. Evolutionary history and adaptation from high-coverage whole-genome sequences of diverse African hunter-gatherers. *Cell* **150**(3):457–469 (2012).

12. Albrechtsen, A., Nielsen, F. C., and Nielsen, R. Ascertainment biases in SNP chips affect measures of population divergence. *Mol. Biol. Evol.* **27**(11):2534–2547 (2010).

13. Brown, F. H., McDougall, I., and Fleagle, J. G. Correlation of the KHS Tuff of the Kibish Formation to volcanic ash layers at other sites, and the age of early *Homo sapiens* (Omo I and Omo II). *J. Hum. Evol.* **63**(4):577–585 (2012).

14. White, T. D., et al. Pleistocene *Homo sapiens* from Middle Awash, Ethiopia. *Nature.* **423**(6941):742–747 (2003).

15. Bräuer, G. The origin of modern anatomy: by speciation or intraspecific evolution? *Evol. Anthropol. Issues, News, Rev.* **17**(1):22–37 (2008).

16. Brauer, G. Middle Pleistocene diversity in Africa and the origin of modern humans. In: Hublin, J.-J., and McPherron, S. P., editors. *Modern origins: A North African perspective.* New York: Springer; 2012. Pp. 221–240.

17. Wood, B. and Leakey, M. The Omo-Turkana Basin fossil hominins and their contribution to our understanding of human evolution in Africa. *Evol. Anthropol.* **20**(6):264–292 (2011).

18. Cann, R., Stoneking, M., and Wilson, A. Mitochondrial DNA and human evolution. *Nature* **325**(6099), 31–36 (1987).

19. Weaver, T. D. Did a discrete event 200,000–100,000 years ago produce modern humans? *J. Hum. Evol.* **63**(1):121–126 (2012).

20. Harris, E. E., and Hey, J. X chromosome evidence for ancient human histories. *Proc. Natl. Acad. Sci. U.S.A.* **96**(6):3320–3324 (1999).

21. Blum, M. G. B., and Jakobsson, M. Deep divergences of human gene trees and models of human origins. *Mol. Biol. Evol.* **28**(2):889–898 (2011).

22. Garrigan, D., et al. Evidence for archaic Asian ancestry on the human X chromosome. *Mol. Biol. Evol.* **22**(2):189–192 (2005).

23. Sjödin, P., et al. Resequencing data provide no evidence for a human bottleneck in Africa during the penultimate glacial period. *Mol. Biol. Evol.* **29**(7):1851–1860 (2012).

24. Li, H., and Durbin, R. Inference of human population history from individual whole-genome sequences. *Nature* **475**(7357):493–496 (2011).

25. Wakeley, J. Metapopulation models for historical inference. *Mol. Ecol.* **13**(4):865–875 (2004).

26. Harding, R. M., and McVean, G. A structured ancestral population for the evolution of modern humans. *Curr. Opin. Genet. Dev.* **14**(6):667–674 (2004).

27. Garrigan, D., and Hammer, M. F. Reconstructing human origins in the genomic era. *Nat. Rev. Genet.* **7**(9):669–680 (2006).

28. Hammer, M. F., et al. Genetic evidence for archaic admixture in Africa. *Proc. Natl. Acad. Sci. U.S.A.* **108**(37):15123–15128 (2011).

29. Rightmire, G. P. Middle and later Pleistocene hominins in Africa and Southwest Asia. *Proc. Natl. Acad. Sci. U.S.A.* **106**(38):16046–16050 (2009).

30. Theunert, C., et al. Inferring the history of population size change from genome-wide SNP data. *Mol. Biol. Evol.* **29**(12):3653–3667 (2012).

31. Harris, K., and Nielsen, R. Inferring demographic history from a spectrum of shared haplotype lengths. *PLoS Genet.* **9**(6):e1003521 (2013).

32. Hoffecker, J. F. The spread of modern humans in Europe. *Proc. Natl. Acad. Sci. U.S.A.* **106**(38):16040–16045 (2009).

33. Cavalli-Sforza, L., Menozzi, P., and Piazza, A. *The history and geography of human genes.* (Princeton, NJ: Princeton Univ. Press; 1994).

34. Jorde, L. The genetic structure of subdivided human population: a review. In: Mielke, J. H. and Crawford, M., editors. *Current developments in anthropological genetics. Vol. 1, Theory and methods.* New York: Plenum; 1980. Pp.135–208.

35. Gutenkunst, R. N., et al. Inferring the joint demographic history of multiple populations from multidimensional SNP frequency data. *PLoS Genet.* **5**(10):e1000695 (2009).

36. Rasmussen, M., et al. An Aboriginal Australian genome reveals separate human dispersals into Asia. *Science* **334**(6052):94–98 (2011).

37. Petraglia, M. D., et al. Out of Africa: New hypotheses and evidence for the dispersal of Homo sapiens along the Indian Ocean rim. *Ann. Hum. Biol.* **37**(3):288–311 (2010).
38. Wall, J. D., et al. A novel DNA sequence database for analyzing human demographic history. *Genome Res.* **18**(8):1354–1361 (2008).
39. Schuster, S. C., et al. Complete Khoisan and Bantu genomes from southern Africa. *Nature* **463**(7283):943–947 (2010).
40. Lohmueller, K. E., et al. Proportionally more deleterious genetic variation in European than in African populations. *Nature* **451**(7181):994–997 (2008).
41. Nelson, M. R., et al. An abundance of rare functional variants in 202 drug target genes sequenced in 14,002 people. *Science* **336**(6090):100–104 (2012).
42. Slatkin, M., and Excoffier, L. Serial founder effects during range expansion: a spatial analog of genetic drift. *Genetics* **191**(1):171–181 (2012).
43. Rasteiro, R., and Chikhi, L. Female and male perspectives on the Neolithic transition in Europe: clues from ancient and modern genetic data. *PLoS One* **8**(4):e60944 (2013).
44. Chikhl, L. Update to Chikhi et al.'s clinal variation in the nuclear DNA of Europeans (1998): Genetic data and storytelling—from archaeogenetics to astrologenetics? *Hum. Biol.* **81**:639–643 (2009).
45. Scally, A., and Durbin, R. Revising the human mutation rate: implications for understanding human evolution. *Nat. Rev. Genet.* **13**(11):745–753 (2012).
46. Klein, R. G. Archeology and the evolution of human behavior. *Evol. Anthropol. Issues, News, Rev.* **9**(1):17–36 (2000).

CHAPTER 7

1. Abecasis, G. R., et al. An integrated map of genetic variation from 1,092 human genomes. *Nature* **491**(7422):56–65 (2012).
2. Smith, J., and Haigh, J. The hitch-hiking effect of a favourable gene. *Genet. Res.* **23**(1):23–35 (1974).
3. Cavalli-Sforza, L., Menozzi, P., and Piazza, A. *The history and geography of human genes.* (Princeton, NJ: Princeton University Press; 1994).
4. Tournamille, C., et al. Disruption of a GATA motif in the Duffy gene promoter abolishes erythroid gene expression in Duffy-negative individuals. *Nat. Genet.* **10**(2):224–228 (1995).
5. Hamblin, M. T., and Di Rienzo, A. Detection of the signature of natural selection in humans: evidence from the Duffy blood group locus. *Am. J. Hum. Genet.* **66**(5):1669–1679 (2000).
6. Hamblin, M. T., Thompson, E. E., and Di Rienzo, A. Complex signatures of natural selection at the Duffy blood group locus. *Am. J. Hum. Genet.* **70**(2):369–383 (2002).
7. Enattah, N. S., et al. Identification of a variant associated with adult-type hypolactasia. *Nat. Genet.* **30**(2):233–237 (2002).
8. Bersaglieri, T., et al. Genetic signatures of strong recent positive selection at the lactase gene. *Am. J. Hum. Genet.* **74**(6):1111–1120 (2004).
9. Copley, M. S., et al. Direct chemical evidence for widespread dairying in prehistoric Britain. *Proc. Natl. Acad. Sci. U.S.A.* **100**(4):1524–1529 (2003).
10. Tishkoff, S. A., et al. Convergent adaptation of human lactase persistence in Africa and Europe. *Nat. Genet.* **39**(1):31–40 (2007).
11. Enattah, N. S., et al. Independent introduction of two lactase-persistence alleles into human populations reflects different history of adaptation to milk culture. *Am. J. Hum. Genet.* **82**(1):57–72 (2008).

12. Wooding, S., et al. Natural selection and molecular evolution in PTC, a bitter-taste receptor gene. *Am. J. Hum. Genet.* **74**(4):637–646 (2004).

13. Campbell, M. C., et al. Evolution of functionally diverse alleles associated with PTC bitter taste sensitivity in Africa. *Mol. Biol. Evol.* **29**(4):1141–1153 (2012).

14. Kim, U., et al. Worldwide haplotype diversity and coding sequence variation at human bitter taste receptor loci. *Hum. Mutat.* **26**(3):199–204 (2005).

15. Fumagalli, M., et al. Signatures of environmental genetic adaptation pinpoint pathogens as the main selective pressure through human evolution. *PLoS Genet.* **7**(11):e1002355 (2011).

16. Grossman, S. R., et al. A composite of multiple signals distinguishes causal variants in regions of positive selection. *Science* **327**(5967):883–886 (2010).

17. Hider, J. L., et al. Exploring signatures of positive selection in pigmentation candidate genes in populations of East Asian ancestry. *BMC Evol. Biol.* **13**(1):150 (2013).

18. Cavalli-Sforza, L. L., and Bodmer, W. *The genetics of human populations.* (New York: W.H. Freeman; 1971).

19. Norton, H. L., et al. Genetic evidence for the convergent evolution of light skin in Europeans and East Asians. *Mol. Biol. Evol.* **24**(3):710–722 (2007).

20. Beleza, S., et al. The timing of pigmentation lightening in Europeans. *Mol. Biol. Evol.* **30**(1):24–35 (2013).

21. Fujimoto, A., et al. A scan for genetic determinants of human hair morphology: EDAR is associated with Asian hair thickness. *Hum. Mol. Genet.* **17**(6):835–843 (2008).

22. Kamberov, Y. G., et al. Modeling recent human evolution in mice by expression of a selected EDAR variant. *Cell* **152**(4):691–702 (2013).

23. Park, J.H. et al. Effects of an Asian-specific nonsynonymous EDAR variant on multiple dental traits. *J. Hum. Genet.* **57**(8):508–14 (2012).

24. Altshuler, D. M., et al. Integrating common and rare genetic variation in diverse human populations. *Nature* **467**(7311):52–58 (2010).

25. Akey, J. M., et al. Population history and natural selection shape patterns of genetic variation in 132 genes. *PLoS Biol.* **2**(10):e286 (2004).

26. Bigham, A., et al. Identifying signatures of natural selection in Tibetan and Andean populations using dense genome scan data. *PLoS Gen.* **6**(9):e1001116 (2010).

27. Xu, S., et al. A genome-wide search for signals of high-altitude adaptation in Tibetans. *Mol. Biol. Evol.* **28**(2):1003–1111 (2011).

28. Beall, C. M. Two routes to functional adaptation: Tibetan and Andean high-altitude natives. *Proc. Natl. Acad. Sci. U.S.A.* **104**(Suppl 1):8655–8660 (2007).

29. Scheinfeldt, L. B., et al. Genetic adaptation to high altitude in the Ethiopian highlands. *Genome Biol.* **13**(1):R1 (2012).

30. Hernandez, R. D., et al. Classic selective sweeps were rare in recent human evolution. *Science* **331**(6019):920–924 (2011).

31. Coop, G., et al. The role of geography in human adaptation. *PLoS Gen.* **5**(6):e1000500 (2009).

32. Pritchard, J. K., Pickrell, J. K., and Coop, G. The genetics of human adaptation: hard sweeps, soft sweeps, and polygenic adaptation. *Curr. Biol.* **20**(4):R208–215 (2010).

33. Pritchard, J. K., and Di Rienzo, A. Adaptation—not by sweeps alone. *Nat. Rev. Genet.* **11**(10):665–667 (2010).

34. Hancock, A. M., et al. Adaptations to climate in candidate genes for common metabolic disorders. *PLoS Genet.* **4**(2):e32 (2008).

35. Glieberman, L. Blood pressure and dietary salt in human population. *Ecol. Food Nutr.* **2**(2):143–156 (1973).

36. Thompson, E. E., et al. CYP3A variation and the evolution of salt-sensitivity variants. *Am. J. Hum. Genet.* **75**(6):1059–1069 (2004).

37. Hancock, A. M., et al. Adaptations to new environments in humans: the role of subtle allele frequency shifts. *Philos. Trans. R. Soc. Lond. B. Biol. Sci.* **365**(1552):2459–2468 (2010).

38. Perry, G. H., et al. Diet and the evolution of human amylase gene copy number variation. *Nat. Genet.* **39**(10):1256–1260 (2007).

39. Akey, J. M. Constructing genomic maps of positive selection in humans: where do we go from here? *Genome Res.* **19**(5):711–722 (2009).

40. Lamason, R. L., et al. SLC24A5, a putative cation exchanger, affects pigmentation in zebrafish and humans. *Science* **310**(5755):1782–1786 (2005).

41. Sabeti, P. C., et al. Genome-wide detection and characterization of positive selection in human populations. *Nature* **449**(7164):913–918 (2007).

CHAPTER 8

1. Duarte, C., et al. The early Upper Paleolithic human skeleton from the Abrigo do Lagar Velho (Portugal) and modern human emergence in Iberia. *Proc. Natl. Acad. Sci. U.S.A.* **96**(3):7604–7609 (1999).

2. Soficaru, A., Dobos, A., and Trinkaus, E. Early modern humans from the Peştera Muierii, Baia de Fier, Romania. *Proc. Natl. Acad. Sci. USA* **103**(46):17196–17201 (2006).

3. Stringer, C., and Andrews, P. Genetic and fossil evidence for the origin of modern humans. *Science* **239**(4845):1263–1268 (1988).

4. Wolpoff, M. Multiregional evolution: The fossil alternative to Eden. In: Mellars, P., and Stringer C.B., editors. *The human revolution: Behavioural and biological perspectives on the origins of modern humans.* Edinburgh, UK: Edinburgh University Press; 1989. Pp. 62–108.

5. Relethford, J. *Genetics and the search for modern human origins.* (New York: Wiley-Liss; 2001).

6. Reich, D., et al. Genetic history of an archaic hominin group from Denisova Cave in Siberia. *Nature* **468**(7327):1053–1060 (2010).

7. Krings, M., et al. Neanderthal DNA sequences and the origin of modern humans. *Cell* **90**(1):19–30 (1997).

8. Noonan, J. P., et al. Sequencing and analysis of Neanderthal genomic DNA. *Science* **314**(5802):1113–1118 (2006).

9. Green, R. E., et al. Analysis of one million base pairs of Neanderthal DNA. *Nature* **444**(7717):330–336 (2006).

10. Wall, J. D., and Kim, S. K. Inconsistencies in Neanderthal genomic DNA sequences. *PLoS Genet.* **3**(10):1862–1866 (2007).

11. Green, R. E., et al. A draft sequence of the Neandertal genome. *Science* **328**(5979):710–722 (2010).

12. Meyer, M., et al. A high-coverage genome sequence from an archaic Denisovan individual. *Science* **338**(6104):222–226 (2012).

13. Prüfer, K., et al. The complete genome sequence of a Neanderthal from the Altai Mountains. *Nature* **505**(7481):43–49 (2014).

14. Klein, R. *The human career: Human biological and cultural origins*. (Chicago: The University of Chicago Press; 1989).

15. Briggs, A. W., et al. Targeted retrieval and analysis of five Neanderthal mtDNA genomes. *Science* **325**(5938):318–321 (2009).

16. Derevianko, A., Shunkov, M., and Volkov, P. A. Paleolithic bracelet from Denisova cave. *Archaeol. Ethnol. Anthropol. Eurasia* **34**(2):13–25 (2008).

17. Dean, D., Hublin, J.-J., Holloway, R., and Ziegler, R. On the phylogenetic position of the pre-Neanderthal specimen from Reilingen, Germany. *J. Hum. Evol.* **34**(5):485–508 (1998).

18. Bermúdez de Castro, J. M., et al. The Atapuerca sites and their contribution to the knowledge of human evolution in Europe. *Evol. Anthropol. Issues, News, Rev.* **13**(1):25–41 (2004).

19. Martinón-Torres, M., et al. Morphological description and comparison of the dental remains from Atapuerca-Sima de los Huesos site (Spain). *J. Hum. Evol.* **62**(1):7–58 (2012).

20. Stringer, C. The status of *Homo heidelbergensis* (Schoetensack 1908). *Evol. Anthropol.* **21**(3):101–107 (2012).

21. Meyer, M., et al. A mitochondrial genome sequence of a hominin from Sima de los Huesos. *Nature* **505**(7483):403–406 (2014).

22. Labuda, D., Zietkiewicz, E., and Yotova, V. Archaic lineages in the history of modern humans. *Genetics* **156**(2):799–808 (2000).

23. Garrigan, D., et al. Deep haplotype divergence and long-range linkage disequilibrium at xp21.1 provide evidence that humans descend from a structured ancestral population. *Genetics* **170**(4):1849–1856 (2005).

24. Garrigan, D., and Hammer, M. F. Reconstructing human origins in the genomic era. *Nat. Rev. Genet.* **7**:(9)669–680 (2006).

25. Shimada, M. K., et al. Divergent haplotypes and human history as revealed in a worldwide survey of X-linked DNA sequence variation. *Mol. Biol. Evol.* **24**(3):687–698 (2007).

26. Lari, M., et al. The microcephalin ancestral allele in a Neanderthal individual. *PLoS One* **5**(5):e10648 (2010).

27. Yotova, V., et al. An X-linked haplotype of Neanderthal origin is present among all non-African populations. *Mol. Biol. Evol.* **28**(7):1957–1962 (2011).

28. Grün, R., and Stringer, C. Tabun revisited: revised ESR chronology and new ESR and U-series analyses of dental material from Tabun C1. *J. Hum. Evol.* **39**(6):601–612 (2000).

29. Coppa, A., et al. Evidence for new Neanderthal teeth in Tabun Cave (Israel) by the application of self-organizing maps (SOMs). *J. Hum. Evol.* **52**(6):601–613 (2007).

30. Grün, R. Direct dating of human fossils. *Am. J. Phys. Anthropol.* **131**(Suppl 43):2–48 (2006).

31. Sankararaman, S., et al. The date of interbreeding between Neanderthals and modern humans. *PLoS Genet.* **8**(10):e1002947 (2012).

32. Shea, J. Transitions or turnovers? Climatically-forced extinctions of *Homo sapiens* and Neanderthals in the east Mediterranean Levant. *Quat. Sci. Rev.* **27**(23–24):2253–2270 (2008).

33. Mellars, P. The impossible coincidence: A single-species model for the origins of modern human behavior in Europe. *Evol. Anthropol. Issues, News, Rev.* **14**(1):12–27 (2005).

34. Wall, J. D., et al. Higher levels of Neanderthal ancestry in East Asians than in Europeans. *Genetics* **194**(1):199–209 (2013).

35. Mellars, P. Neanderthals and the modern human colonization of Europe. *Nature* **432**(7016):461–465 (2004).

36. Klein, R. *The human career: Human biological and cultural origins.* (Chicago: Univ. of Chicago Press; 2009).

37. Finlayson, C., et al. Late survival of Neanderthals at the southernmost extreme of Europe. *Nature* **443**(7113):850–853 (2006).

38. Wood, R. E., et al. Radiocarbon dating casts doubt on the late chronology of the Middle to Upper Palaeolithic transition in southern Iberia. *Proc. Natl. Acad. Sci. U.S.A.* **110**(8):2781–2786 (2013).

39. Longo, L., et al. Did Neanderthals and anatomically modern humans coexist in northern Italy during the late MIS 3? *Quat. Int.* **259**(9):102–112 (2012).

40. Fu, Q., et al. A revised timescale for human evolution based on ancient mitochondrial genomes. *Curr. Biol.* **23**(7):553–559 (2013).

41. Sánchez-Quinto, F., et al. Genomic affinities of two 7,000-year-old Iberian hunter-gatherers. *Curr. Biol.* **22**(16):1494–1499 (2012).

42. Haak, W., et al. Ancient DNA from European early Neolithic farmers reveals their Near Eastern affinities. *PLoS Biol.* **8**(11):e1000536 (2010).

43. Rasmussen, M., et al. An Aboriginal Australian genome reveals separate human dispersals into Asia. *Science* **334**(6052):94–98 (2011).

44. Reich, D., et al. Denisova admixture and the first modern human dispersals into Southeast Asia and Oceania. *Am. J. Hum. Genet.* **89**(4):516–528 (2011).

45. Currat, M., and Excoffier, L. Strong reproductive isolation between humans and Neanderthals inferred from observed patterns of introgression. *Proc. Natl. Acad. Sci. U.S.A.* **108**(37):15129–15134 (2011).

46. Jolly, C. J. A proper study for mankind: Analogies from the papionin monkeys and their implications for human evolution. *Am. J. Phys. Anthropol.* **116**(Suppl 33):177–204 (2001).

47. Mendez, F. L., Watkins, J. C., and Hammer, M. F. Neanderthal origin of genetic variation at the cluster of OAS immunity genes. *Mol. Biol. Evol.* **30**(4):798–801 (2013).

48. Mendez, F. L., Watkins, J. C., and Hammer, M. F. A haplotype at STAT2 introgressed from Neanderthals and serves as a candidate of positive selection in Papua New Guinea. *Am. J. Hum. Genet.* **91**(2):265–274 (2012).

49. Abi-Rached, L., et al. The shaping of modern human immune systems by multiregional admixture with archaic humans. *Science* **334**(6052):89–94 (2011).

50. Kelaita, M. A., and Cortés-Ortiz, L. Morphological variation of genetically confirmed Alouatta Pigra × A. palliata hybrids from a natural hybrid zone in Tabasco, Mexico. *Am. J. Phys. Anthropol.* **150**(2):223–234 (2013).

51. Gunz, P., and Harvati, K. Integration and homology of "Chignon" and "Hemibun" morphology. In: Condemi, S. and Weniger, G.-C., editors. *Continuity and discontinuity in the peopling of Europe.* New York: Springer; 2011. Pp. 193–202.

52. Tattersall, I., and Schwartz, J. H. Commentary: Hominids and hybrids: the place of Neanderthals in human evolution. *Proc. Natl. Acad. Sci. U.S.A.* **96**(13):7117–7119 (1999).

53. Krause, J., et al. The derived *FOXP2* variant of modern humans was shared with Neanderthals. *Curr. Biol.* **17**(21):1908–10912 (2007).

54. Maricic, T., et al. A recent evolutionary change affects a regulatory element in the human *FOXP2* gene. *Mol. Biol. Evol.* **30**(4):844–852 (2013).

55. Cerqueira, C. C. S., et al. Predicting *Homo* pigmentation phenotype through genomic data: from Neanderthal to James Watson. *Am. J. Hum. Biol.* **24**(5):705–709 (2012).

56. Lalueza-Fox, C., et al. A melanocortin 1 receptor allele suggests varying pigmentation among Neanderthals. *Science* **318**(5855):1453–1455 (2007).

EPILOGUE

1. Klein, R. *The human career: Human biological and cultural origins* (Chicago: Univ. of Chicago Press; 2009).

INDEX

Chen, Feng-Chi, 41, 43

Chewing muscles and expansion of brain size, 78, 96

Chikhi, Lounès, 132

Chimpanzees: ABBA-BABA test, 173; adaptive change and, 87, 89(figure), 90, 91–92; brain of, 92–93, 94, 96, 98, 100–101, 110; female promiscuity in, 90; genome of, xv, 15, 44, 56–57, 87, 114; human diseases not affecting, 60, 77, 79; interbreeding with humans (hypothesis), 52–53; lice in, 103; pelvis of, 109; relationship to gorillas, 38; relationship to humans, xvii, 13, 15, 18, 20, 21, 22, 32, 33, 35, 191; skin pigmentation of, 102; speciation process (see Speciation)

Chromosome(s): crossing over, 25–26, 32, 172; defined, 195; effective population size and, 61; homology and, 10; number of in great apes, 25; number of in humans, 25; X, 62–63, 123–24; Y, 113–14, 132

Chromosome 1, 33, 34(figure)

Chromosome 11, 21

"Chumanzees," 52

Cichlid fish, 22

Clades, 6

Cladistics, 6, 16–17

Coalescence, 24–25, 28–31; defined, 24, 195; human origins and, 123–24; speciation and, 36–38, 45–48, 50, 57

Cochlea, 107

COII gene, 20, 21

Collard, Mark, 11

Color blindness, 66

Condemi, Silvana, 180

Continuous variation, 154

Coronary artery disease, 77

Cortés-Ortiz, Liliana, 185

Cranium, shape of, 117

Creatine, 98

Crick, Francis, xv

Crohn's disease, 60, 76–77

Crossing over, 25–26, 32, 172

CRYGD gene, 79

CYP3A5 gene, 156

Cystic fibrosis, 130

Cytosine (C), xvi, 8

Dai Chinese, 178

DARC gene, 140–42, 144, 146, 148, 150, 152, 160

Dart, Raymond, xviii

Darwin, Charles, xvii–xviii, xx, 14–15, 20, 27, 34–35, 67, 85, 86, 92, 94, 111, 194

Darwin's finches, 109

Dausset, Jean B., 118

Dementia, 79

Dendrites, 94

Denisova Cave (Siberia), 164, 165(figure), 170–71, 175, 180

Denisovans, 170–71, 172, 181; discovery of, 164–65; FOXP2 gene in, 187; genome of, xix, 165, 168, 184, 192–93; interbreeding with modern humans, 165, 177–78, 182, 183, 184, 185, 193; interbreeding with Neandertals, 189; mitochondrial genome of, 167; nuclear genome of, 167; skin pigmentation of, 188; split from Neandertals, 170

De novo (pedigree) mutation rate, 133–35, 169

Descent of Man, The (Darwin), xvii–xviii, 14–15, 111

Diabetes: type 1, 77; type 2, 60, 77, 156

Diet, 157–59

Di Rienzo, Anna, 141–42, 157

Disease: age-related, 98–99; effective population size and, 60, 76–78, 79; gene surfing and, 130–31; negative selection and, 76–78; polygenic, 77; population bottleneck and, 129–30; species-wide adaptations and, 88

Disotell, Todd, 3–4, 10, 11

Divergence, 40, 47, 103, 148, 149, 195

Diversity Project, 118

DNA: amino acid-altering sites (see Amino acid-altering sites/mutations); ancient (see Ancient DNA); autosomal, 123–25, 195; chimpanzee, 42, 44, 47, 87, 91–92; Denisovan, 170, 178, 181, 185; discovery of molecular structure, xv; effective population size and, 61–63; gorilla, 42, 44, 47; HGDP-CEPH collection, 118–19; homologous, 7–10, 25; junk, xvi, 99–100; mitochondrial (see

DNA: amino acid-altering sites (*Cont.*)
Mitochondrial DNA); Neandertal,
170, 173–75, 180–81, 185, 190;
nuclear, 63, 113–14, 122–23, 196;
orangutan, 33–34, 43; silent sites,
72–73, 88, 103; speciation and,
41–43, 47, 50, 57–59; species-wide
adaptations and, 87, 88; "suitcases"
of, 178, 181; tag-alongs in site,
113–14
DNA chip, 119–20
DNA-DNA hybridization, 15–17
DNA sequences, 3–4, 8, 9, 10, 17–21,
31, 117, 122, 137, 138, 140, 141,
145, 147, 167, 182; comparison of in
SNPs, 115(figure); defined, 17; of gene
regulatory regions, 100; largest study
of, 63. *See also* Genome sequencing;
1000 Genomes Project
DNA variants: in brain genes, 93–94;
defined, 195; disease and, 60, 77,
79; effective population size and, 60,
65–66, 68, 69, 77, 79, 80, 83–84;
gene surfing and, 130; hair texture
and, 149, 150; height variation and,
154–55; lactose tolerance and, 142–
44; malaria and, 141–42; microarray
studies of, 119–20; natural selection
and, 137–40; population bottleneck
and, 124, 129–30; population-specific
adaptations and, 137–40, 145,
146, 149, 150, 152, 154–55; skin
pigmentation and, 188
Dolphins, 7, 92, 95–96, 97(figure), 111
Dopamine, 106
Drift. *See* Genetic drift
Dunkers, 64, 66–67
Durbin, Richard, 134

Earlobes, unattached, 66–67
East Asia: colonizing of, 129; Neandertal
DNA in population, 180
EDAR gene, 149–50
Effective population size, 39–44,
60–84, 128; of ancestral population,
40–44, 45, 60; defined, 39, 195; and
relationship to census population
size, 39, 82, 128; downside of crash,
73–78; estimating, 61–64; genetic
consequences of small, 64–67; genetic

drift and, 62, 65–70, 71, 80, 81, 83; of
modern humans, 39, 44, 45, 60, 63,
170; natural selection and (*see under*
Natural selection; Negative selection;
Positive selection); of Neandertals,
170; upside of crash, 78–82
Elephants, 95–96, 97(figure), 111
Ellis-van Creveld syndrome, 67
El Sidrón Cave (Spain), 170, 187, 188
Enard, Wolfgang, 106
Encephalization quotient (EQ), 92
ENCODE project (Encyclopedia of DNA
Elements), 100
Endometriosis, 77
Endurance running while hunting, 104
Enlarged heart, 79
Environmental correlation, 156
Epistasis, 79, 195
Epithelial cancer, 77
EQ. *See* Encephalization quotient
Eurasia: ancient DNA from, 126–27;
colonizing of, 130, 131, 134, 136;
Neandertal DNA in population, 175,
190; split from African population,
128–29
Europe: agricultural lifestyle in, 132;
colonizing of, 129; DNA variants
underlying diseases in, 130; effective
population size in, 62; lactose tolerance
in population, 142, 144; Neandertal
DNA in population, 185; Neandertal
habitation in, 163, 180–81; skin
pigmentation in population, 149, 160
Evolutionary convergence, 7, 28, 144,
149
Evolutionary trees, 2, 14(figure);
close-up view of, 163(figure); fossils'
place on, 12; homology and, 5–10;
how to grow, 16–17; Neantertals
on, 162–63; papionin monkeys on,
4, 5(figure), 11. *See also* Gene trees;
Species trees
Exome, 76
Expensive tissue hypothesis, 95
EYA1 gene, 108
Eyre-Walker, Adam, 81

Falciparum malaria, 77, 118, 141
Family trees. *See* Evolutionary trees
Female migration, 132

Human genome: chimpanzee genome compared with, 44, 87, 114; cloudy evolutionary history of, 32–35; length of, xvi; number of genes in, xvii, 23; percentage of genes in, xvi, 99; sequencing of, xv, 23, 191; similarity across species, xvi; sticky problem, 15. *See also* 1000 Genomes Project

Humanized mice, 106–7

Human leukocyte antigen (HLA) genes, 184–85

"Human revolution" hypothesis, 134

Humans. *See* Archaic hominins; Modern humans

Hunter-gatherer groups, 119, 120, 123, 131, 132, 180; ancient mitochondrial lineages of, 122; archaic DNA in genomes, 127; dietary specializations in, 158

Hutterites, 64, 66, 67, 68

Hybrid zones, 181–85; as evolution's laboratories, 181–182

Hypertension, 77, 156

Iberia, 179

Immune genes, 184–85

Immunoglobulin C alpha 1 gene, 20

Inbreeding, 64–67

India, 134

Industrial Revolution, 39

Infectious disease. *See* Pathogens

Informative sites, 18

In-group species, 18

Interbreeding: of Eurasian and African populations, 128–29; human-chimpanzee (hypothesis), 52–53; hybrid zones, 181–85; modern human-archaic hominin, xix, 126–28, 133, 172–81, 192; modern human-Denisovan, 165, 177–78, 182, 183, 184, 185, 193; modern human-Neandertal, 161–64, 172–77, 178–81, 182, 183, 184, 185, 190, 193; Neandertal-Denisovan, 189; of nonhuman primate species, 49–50; possible benefits of, 183–85

Internode, 31, 40–42, 43; defined, 30; determining, 40

Intron regions, 100

Involucrin gene, 20

Island of the Colorblind and Cycad Island (Sacks), 66

Isolates, 64–67, 68, 71

"Isolation by distance" pattern, 115

Japanese people, 158

Jaw, decreased size of, 78, 96, 108, 109, 161

Jemez tribe, 67

Jolly, Clifford, 11, 49, 182

Junk DNA, xvi, 99–100

Kappelman, John, 43

Kebara site (Israel), 175

Kelaita, Mary, 185

King, Mary-Claire, 99, 107

Klein, Richard, 134

Kluge, Arnold, 15

Labuda, Damian, 172, 173

Lactase gene, 100, 142–44, 153, 160

Lactose tolerance, 142–44, 149, 160

Lagar Velho. *See* Old Mill Rock Shelter

Lalueza-Fox, Carles, 187, 188–89

Language, 85, 86, 88, 91, 92, 105–8, 110, 111; *FOXP2* gene in (*see FOXP2* gene); Neandertal question, 105, 186–88; time of acquisition of, 105

Lemurs, 95

Lengkieki (typhoon), 66

Lesser hedgehog tenrec, 96, 97(figure)

"Less is more" hypothesis, 78

Li, Wen-Hsiung, 41, 43

Lice, 103–4

Lieberman, Daniel, 60, 104

Linkage disequilibrium, 115–16

Linkage groups, 26, 32, 33, 37, 138, 196

Linnaeus, Carolus, 16

"Lucy" skeleton, 12

Lysozyme, 27–28

Macaque monkeys, xv, 49

Maddison, Wayne, 33

Malaria: candidate gene studies of, 140–42; falciparum, 77, 118, 141; vivax, 141–42, 150

Mammalian clade, 6

Mandrills, 2, 9, 10, 12

Mangabey monkeys, 2, 3–4, 10, 12

Mating systems, 90–91. *See also* Inbreeding; Interbreeding

MATP gene, 173

Max Planck Institute, 106, 165

Mayr, Ernst, 49

McGraw, Scott, 11

MCPH1 gene, 93–94

MC1R gene, 91, 102–3, 188–89

Meat eating, 96–99

Melanesia, 69–70, 177, 184

Melanosomes, 160

Mellars, Paul, 176, 179

Mennonites, 64

Messenger RNA. *See* mRNA

Metapopulation model of modern human origins and early evolution, 124–26

Mexican cavefish, 8

Mexico, 182

Mezmaiskaya Cave (Caucasus), 168, 170

Mezzena Shelter (Italy), 180

Mice: coat color significance, 8, 91; *EDAR* gene inserted in, 150; Gulf Coast and Atlantic beach, 8; humanized, 106–7; negative selection and, 74–75, 77; positive selection and, 81–82; vocalization in, 106–7

Microarray studies, 119–20

Microcephaly, 93–94, 108

Microencephalin gene, 173

Microsatellites, 117, 119

Milk, digestion of. *See* Lactose tolerance

"Millenium Man" (*Orrorin* nickname), 54

Mineral replacement, 166

Mitochondrial DNA (mtDNA), 5 (figure), 10, 62, 112, 113, 114, 124, 132, 167, 196

Mitochondrial genomes, 3–4, 20, 32; Denisovan, 167; effective population size and, 62–63; in human origin studies, 118, 132; inheritance of, 3; length of, 112; limitation of studies based on, 113–14; Neandertal, 166–67; of Neandertals' direct ancestors, 171–72; nuclear genome compared with, 62–63; population bottleneck support in, 128; pullouts of, 4(figure)

Mitochondrial trees, 5(figure), 122–23

Modern humans: defined, 196; effective population size, 39, 44, 45, 60, 63, 170; expansion out of Africa, 128–31, 134–35, 136; interbreeding with archaic hominins (*see under* Interbreeding); origins of, 62, 114–20; rate of emergence, 120–26; split from archaic hominins, 168–72

Molecular clock-based method, 42, 62

Molecules versus Morphology: Conflict or Compromise? (Patterson), 1

Monkeys: African, 94; brain of, 94, 95; howler, 182, 185; macaque, xv, 49; mangabey, 2, 3–4, 10, 12; New World, 7; Old World, 2, 8; papionin, 2–5, 10–12; rhesus, 74; South American, 94; spider, 7

Morphology: genetics and, 1–5, 10–12; homology compared with, 8–9

Morris, Desmond, 101

Mostly Out of Africa hypothesis, 164

mRNA, 100, 110

MTRR gene, 157

Multiregional Continuity model, 163–64

Music, 108

Musical instruments, 108

Mutations: amino acid-altering (*see* Amino acid-altering sites/mutations); *de novo* (pedigree) rate, 133–35, 169; effective population size and, 61–62; genetic hitchhiking of neutral, 138–40; of microsatellites, 117; of mitochondrial genome, 20; molecular theory on, 18; of nuclear genome, 20; phylogenetic rate, 43, 133–34, 169; rate of in humans *vs.* chimpanzees, 56; silent, 72–73, 88, 103

Mwahuele, Nahnmwarki, 66

Myelination, 101

MYH16 gene, 78, 96

Myocardial infarction, 77

Myosin, 78

Naked Ape, The (Morris), 101

Native Americans, 67, 150

Natural selection: balancing, 145, 172, 184–85; cranium shape not molded by, 117; defined, 196; effective population size and, 67–70, 71–73, 82–84; genetic drift and, 67–70, 71, 81, 83; gene trees and, 27–28, 113; guide to, 137–40; height variation and, 154–56; in hybrid zones, 182–83; negative (*see* Negative selection); population-specific adaptations and, 136–40, 154–56; positive (*see* Positive selection); rebound of, 82–84; species-wide adaptations and, xviii–xix, 85

Navajo tribe, 67

Nazlet Khater (Egypt), 127

Neandertals, 133, 178; effective population size, 170; emergence of lineage, 171–72; first discoveries of, xix; genome of, xix, 168, 179, 186–88, 192; geographic distribution of, 175–77; interbreeding with Denisovans, 189; interbreeding with modern humans, 161–64, 172–77, 178–81, 182, 183, 184, 185, 190, 193; language question, 105, 186–88; mitochondrial genome of, 166–67; nuclear genome of, 167; skin pigmentation of, 188–89; split from Denisovans, 170; split from human lineage, 169–70

Neander Valley site (Germany), 167

Negative selection: defined, 68; effective population size and, 68, 70, 71, 72–78, 83; gene surfing and, 130; genetic drift and, 68, 70; measuring effectiveness of, 73–78; population bottleneck and, 129–30; purifying selection as alternative term, 68, 197; skin pigmentation and, 102

Nei, Masatoshi, 21, 41

Neocortex, 100

Neoteny, 101

Nesse, Randolph, 60

Neurogenomics, 93

Neurons, 94, 101

Neutrality, 73

New World monkeys, 7

Nielsen, Rasmus, 119

Non-tasters, 145

Nuclear DNA, 63, 113–14, 122–23, 196

Nuclear genomes, 20; benefits of using for studies, 113–14; Denisovan, 167; effective population size and, 62–64; in human origin studies, 114–20, 132; mitochondrial genome compared with,

62–63; Neandertal, 167; population bottleneck support in, 128; pullouts of, 4(figure)

Nucleotides: defined, xvi; homology of, 8, 9

"Nutcracker Man" (*Paranthropus boisei* nickname), 12

OAS1 gene, 184

Obesity, 77, 156

Occam's razor, 9

Occipital bun, 162, 185–86

Ohno, Sasumu, 99

Old Mill Rock Shelter (Portugal), 161–62, 164, 185, 186

Old World apes, 2

Old World monkeys, 2, 8

Olfaction, 77, 78, 82, 88, 90, 148

Olorgesailie site (Kenya), 2

Olson, Maynard, 78

Omo Kibish site (Ethiopia), 121, 122, 124

Omo I skull, 122

Omo II skull, 122

1000 Genomes Project, 74, 120, 137, 152, 179, 195

On the Origin of Species (Darwin), 34, 67, 85, 194

Orangutans: Asian, 13; genome of, xv, 45; relationship to humans, 13, 14, 15, 18, 21, 33–34; speciation process, 41, 43, 45, 47, 52, 169

Oreopithecus, 55–56

Orrorin, 54, 56

Out-group species, 18

Outlier genes, 148, 159, 160

"Out of Africa" model, xix, 122, 123, 129, 130, 163, 164

"Out of Africa" population bottleneck, 123, 128-130

Pääbo, Svante, 165, 166–68, 169, 170, 171, 187

Paisa, 64

Paleoanthropology, 12

Paleogenomics, 164, 166, 190

Pamilo, Pekka, 21, 41

Papuans, 169

Pan, 18

Papionin monkeys, 2–5, 10–12

Papua New Guinea, 177

Sabeti, Pardis, 149–50
Sacks, Oliver, 66
Sahelanthropithecus, 56
Sahelanthropus, 54
Sahelanthropus tchadensis, 58
Salt-sensitive hypertension, 156
San Bushmen, 123, 127, 169
Sandawe, 120, 127
Sanger Institute, 134
Sapolsky, Robert, 98
Savanna hypothesis, xvii–xviii, 52–53
Scally, Aylwyn, 134
Schwestergruppen, 16
SCN5A gene, 79
Seesaw metaphor for balance between
    genetic drift and natural selection,
    67–68
Seipel, Adam, 73
Selective sweep, 113–14, 138–40, 145,
    152–53, 156, 160, 192; defined, 197;
    of *FOXP2,* 187; hard sweep, 153;
    lactase, 142; malaria, 141–42; skin
    pigmentation, 149; soft sweep, 153,
    156, 159, 160
Sepsis, 78
Serial founder effect, 115–19, 128
Sexual reproduction, 25–26. *See also*
    Inbreeding; Interbreeding
Shanidar Cave (Iraq), 175
Shared derived features, 16, 17
Shovel-shaped incisors in Asians and the
    EDAR gene variant, 150
Siamangs, 13
Sibley, Charles, 15
Siepel, Adam, 63
Signature of selection, 140, 145–46,
    153; genome scanning for, 146–48;
    distinguishing from signature due
    to demographic history, 146, 147,
    147(figure); for lactose tolerance, 142;
    limitations to detecting signature of
    selection, 91
Silent sites/mutations, 72–73, 88, 103
Sima de los Huesos. *See* Pit of Bones
Single nucleotide polymorphisms
    (SNPs), 114–17, 140, 146, 150–52,
    159, 160; comparisons of DNA
    sequences in, 115(figure); defined,
    197; dietary specializations and,
    157; in pigmentation and

metabolic genes, 156. *See also*
    Perlegen Project
Sivapithecus, 43
Skhul Cave (Israel), 134–35, 175–76,
    177(figure)
Skin pigmentation, 101–3, 104, 148–49,
    156, 160, 188–89
SLC24A5 gene, 160
Smell, sense of. *See* Olfaction
Smith, John Maynard, 137–38
SNPs. *See* Single nucleotide
    polymorphisms
Sodium-retention hypothesis, 156
Soft sweep, 153, 156, 159, 160
South American Indians, 178
South American monkeys, 94
South Asian Islanders, 178
Southeast Asia: colonizing of, 129, 134;
    Denisovan DNA in population, 181
Speciation, 28–31, 36–59, 103, 112,
    169; allopatric, 49, 50, 51, 195;
    ancestral imposters, 53–59; dates of,
    41–43, 45–50, 54, 56, 91; defined,
    22, 197; parapatric, 50, 51(figure),
    197; prolonged process hypothesized,
    51–53; proposed scenarios of, 49–50;
    savanna hypothesis, xvii–xviii, 52–53;
    sympatric, 50; bottleneck (modern
    humans) 123–124
Species: defined, 197; molecular
    theory on differences in, 18. *See also*
    individual species
Species tree-gene tree congruence, 26
Species trees, 26–27; defined, 197;
    explained, 22–23; as "fat trees,"
    26, 27, 57; for papionin monkeys,
    3(figure). *See also* Gene tree-species
    tree incongruence; Gene tree-species
    tree mismatch; Gene tree-species tree
    problem
Species-wide adaptations, xviii–xx,
    85–111, 140; genome-scanning
    studies of, 86–91, 92, 93, 107–8, 109,
    148; limitations to detecting, 91–92;
    successes and challenges of studies,
    109–11
Spermatogenesis, 88, 90, 148
Sperm Competition, 90
Spider monkeys, 7
Stanford, Craig, 98